Specialty Food Packaging Design

Specialty Food Packaging Design

by Reynaldo Alejandro

with the

Distributor to the book trade in the United States and Canada:
Rizzoli International Publications, Inc.
597 Fifth Avenue
New York, NY 10017

Distributor to the art trade in the United States:
Letraset USA
40 Eisenhower Drive
Paramus, NJ 07653

Distributor to the art trade in Canada:
Letraset Canada Limited
555 Alden Road
Markham, Ontario L3R 3L5, Canada

Distributed throughout the rest of the world by:
Hearst Books International
105 Madison Avenue
New York, NY 10016

Library of Congress Cataloging-in-Publication Data

Alejandro, Reynaldo G.
Specialty food packaging design / by Reynaldo Alejandro.
p. cm.
Includes index.
ISBN 0-86636-105-7
1. Packaging—Design. 2. Food—Packaging. I. Title.
TS195.4.A44 1989
664'.092—dc20 89-9225
CIP

Color separation, printing, and binding by
Toppan Printing Co. (H.K.) Ltd. Hong Kong
Typesetting by ACS Communications, Inc.

MANAGING EDITOR: KEVIN CLARK
EDITOR: JOSEPH DIONISIO
ASSOCIATE ART DIRECTOR: DANNY KOUW
ARTISTS: KIM McCORMICK
CARLO BONOAN

ACKNOWLEDGEMENTS

I would like to thank the many people who made this book possible—from PBC International, Inc: Penny Sibal-Samonte, my ever-inspired Muse; Richard Liu; Daniel Kouw; Joseph Dionisio; Lawrence Schimel; and the ever-helpful editor who edited my first cookbook and my *Classic Menu Design* book, Kevin Clark.

For assistance in various areas of research I am very indebted to Cielo Fuentebella, Roberto Labog Tanizaki, Willie Barranda, Joseba Encabo, Luis Guevara, Miguel Angel Martinez Alonzo, Marco Maltos, and Cesar Bernardo.

And above all to the late Herb and Cora Sibal-Taylor who always have inspired me, may their souls rest in peace.

CONTENTS

TryMe®
ZESTY AND FULL OF FLAVOR
HOT PEPPER
TENNESSEE SUNSHINE®
SAUCE
GREAT TASTE FOR BURGERS AND BEEF
BULLFIGHTER
STEAK AND BURGER SAUCE
5 fl. oz. (147 ml.)
REAL CAJUN FLAVOR
HOT PEPPER SAUCE
CAJUN SUNSHINE®
SPICY AND DELICIOUS ON SEAFOOD
SAUCE
A SPICY SAUCE FOR MEATS & SEAFOOD
TIGER
SAUCE®
THE ORIGINAL
WITH FINE SHERRY AND PEPPERS
Wine & Pepper
FANCY
WORCESTERSHIRE SAUCE

FOREWORD

A Perfect Combination of Beauty and Taste

The specialty food industry in the U.S. has thrived by developing top-quality products, and presenting them in packaging that appeals to the aesthetic tastes of sophisticated consumers.

In the March/April 1989 issue of NASFT Showcase, the National Association for the Specialty Food Trade, Inc., asked manufacturers/importers/distributors and retailers to rank 12 factors that influence whether consumers will purchase new specialty food products. Both groups ranked packaging as the most important factor, followed by the uniqueness of the product.

In an effort to match and exceed the packaging beauty of European specialty foods, American manufacturers have greatly upgraded the appearance of their products. Today, attractively packaged European imports and American-produced foods vie for the attention of specialty food shoppers.

It makes economic sense for specialty food manufacturers to devote energy and money to producing beautiful packaging. Since most of these foods are produced in a limited quantity, attention to detail is the rule. Unlike mass-produced foods, these items are not coming off factory production lines in which the time and cost of production are the ruling factors. Rather, most specialty foods are produced and packaged in small production facilities or by hand. Close attention can then be paid to the beauty of the packaging as well as the quality of the food.

In addition, the egos of the entrepreneurs who create these foods demand that the quality of the package reflects the quality of the product. Specialty food manufacturers and importers are proud of their products, and want them presented in the best possible way.

These entrepreneurs do not have the funds to hire the big-name package designers who create products for major food corporations. They either have the work done by unknown yet highly creative designers, or they do it themselves. The creativity that they put into their product is reflected in the ingenuity of their packaging.

As you can see in this beautiful book, the packaging developed by U.S. specialty food manufacturers and importers is truly exquisite. From Italian-designed packages that hold tasty chocolates to eye-catching graphics developed by food producers with an eye for the bold, the packages that contain products sold by the 1,200 members of the National Association for the Specialty Food Trade, Inc., are truly works of art. That will be evident as you peruse the hundreds of photographs published in this beautiful book.

RONALD TANNER
NASFT Communications Director

INTRODUCTION

No longer the purchases of the well-to-do, the specialty food industry is booming.

"Fancy foods are really mainstream today," says Mona Doyle of The Consumer Network, which surveys more than 4,500 consumers about their product preferences every month. "Specialty foods are a piece of the good life almost everyone can afford. The appreciation and recognition of specialty food is growing and will continue to grow." *

The rise of money-laden, young, urban professional consumers and an influx of exotic and unique cuisine has created a marketplace for gourmet and unusual food items.

Gourmet bake shops, specialty coffee and tea emporiums, confectionery stores and high-end, quality-conscious supermarkets are booming. Today's consumer demands quality, taste and health, and is willing to pay for it.

Contained within these pages are examples of some of the premium specialty food items available today. These select food products were chosen for their unique packaging. Other criteria included: esthetic value, ease of display, interplay between package and product, logoture, graphics, illustration, packaging material, new production technology and ease of use.

Welcome to the world of specialty food; we hope you enjoy this book as much as we do.

** As appeared in the March/April 1989 edition of NASFT Showcase.*

Chapter One

FINE CANDIES CHOCOLATES

Chocolate, one of life's sweetest pleasures, is an addiction almost impossible to break. "But the good news," according to chocolate expert Joan Steur of New York, Chocolate Marketing Inc., "is chocolate is not 'bad' for your health. Chocolate," Steur said in an interview with *Gourmet Today*, "is low in sodium, does not contain caffeine, does not promote acne or tooth decay and is not fattening in small amounts."

Statistics show that consumers are not giving up chocolate despite dire predictions from medical professionals regarding the sugar and cholesterol content. According to a study commissioned by the National Confectioners Association, the National Candy Wholesalers Association and the Chocolate Manufacturers Association, confectionery sales have increased 43 percent since 1982. Nine out of ten U.S. households buy confections, while 56 percent buy and eat candies every week.

Chocolate represents just over half of the $8 billion U.S. candy industry. In the next five years, most of us will consume over 60 pounds of chocolate. "People want to indulge, both for themselves and for gifts," said Van Billington, executive director of Retail Confectioners International, whose membership consists of some 600 manufacturers/retailers.

Despite its stable position in the market, price, packaging and taste are challenges for retailers. Diet-conscious consumers, meanwhile, are looking for chocolate low in sugar, fat and cholesterol.

Health concerns have changed Americans' perceptions regarding confections. Few consumers consider candy a "healthful snack" and eight out of 10 surveyed say that candy contains more sugar than is good for them. To satisfy the sweet tooth of consumers who feel guilty about eating regular confections, several companies now offer sugar-free and low-fat candies.

Other manufacturers have reduced the butterfat and sugar content of their candies, so a diet truffle contains 40 calories, versus 80 to 100 calories for a conventional version. Sugar-free candies and chocolates are not necessarily low in calories, as some customers mistakenly presume. Many sugar-free confections sweetened with mannitol or other sugar-free products are also heavy in fat from nuts and rich chocolate.

Chocolate has come a long way since the Aztecs introduced the *xacolati* brew in the 1500s to Spanish explorers. In 1756, it was first manufactured in the United States, now the world's leading producer. Today, gourmet chocolate stores sell dozens of flavors by the piece or by the pound. Old and new recipes are spurring product growth as both imported and domestic brands vie for market share.

Smaller-sized confections and creative designs are also in fashion. Many prestigious manufacturers favor small moulded chocolates over the larger-sized bars and candies. As one New York retailer says, "People want to treat themselves to something that's rich. A truffle the size of a golf ball is too much of a good thing — then it's not a treat, it's a full dessert."

One retail outlet has a product line that comes in the shape of small seashells, decorated in hues of white, milk and dark chocolate. Truffles are now packaged in symmetrical mounds often topped with pastel swirls, while novelties can be anything from chocolate greeting cards to figurines. But the smart money is on chocolate creations such as long stemmed roses, oversized kisses, angel figures, golf bags, license plates and lipstick tubes.

Current best sellers include truffles, chocolate-dipped potato chips and pretzels, fruits, nuts, Turtles, sugar-free and low-sugar products, chocolate bark, novelties, upscale chocolate bars, fudge and chocolate sauces. The popularity of macadamia nuts, pecans and almonds—all homegrown items—contribute to the appeal of barks and brittles. Market analysts predict that other American classics such as marshmallow cups, peanut butter cremes and fruit fillings will be future sizzlers.

Among imported brands, retailers say that Italian and Swiss confections are getting the most attention and business. Truffles, liqueur fillings and white chocolate are among the chief European influences on the U.S. market. While most state laws still allow no more than 0.5 percent liquor content in confections, some states are lifting restrictions.

As imports of liqueur-filled chocolates gain popularity, domestic manufacturers are making plans for their own product lines. Liqueur-filled candies give consumers a chance to be risque without becoming intoxicated. Since the cost of fine imported chocolates has risen in proportion to U.S. dollar's decline, some retailers are bringing chocolate production in-house.

Another noticeable consumer trend is the movement away from milk chocolate—long the American preference. As consumers become more sophisticated, white chocolate from European chocolatiers and dark bittersweet chocolates are becoming a mainstream taste. Outside of metropolitan areas, however, retailers say that milk chocolate is still the preferred choice.

Equally significant is the trend toward sugar-free chocolate. "Chocolate is an addictive food," said Bob MacLachlan, president of the consulting firm, H.M. International, in an interview with *Fancy Food.* "It has caffeine and phenylethymine which can give the nervous system a kick and elevate mood." The demand for sucrose-free chocolate is increasing, particularly among diabetics and people with dental concerns.

Recent studies at the Forsyth Dental Center in Boston, however, cite an ingredient in chocolate as a barrier to tooth decay. Chocolatiers can thus appeal to the safe pleasures of eating chocolates. Still, a number of manufacturers are producing "healthy" chocolates that are either sugar-free, dairy-free, or cholesterol-free. Tofu and cocoa butter, a highly saturated fat, are often used in "chocolate for the healthy gourmet." Fat content, however, is the biggest stumbling block in producing low calorie chocolates. At an average of 150 calories and 11 grams of fat per ounce, even sugarless chocolate is fattening.

As the chocolate and confectionery market becomes more competitive, retailers are facing customers who are conscious of quality and are willing to pay for it. Packaging is another factor that influences purchasing decisions.

"Packaging is something you can enhance or even create at your store." A Michigan-based chocolatier explains, "Packaging plays a tremendous role in sales and this is where we can offer originality. For a special gift, we'll stack three or four boxes on top of each other and put it together with beautifully-colored ribbons and bows."

Other retailers custom-package their goodies in baskets, tins and mugs. For real "chocoholics," retailers may wish to design a "chocolate" basket with multi-colored tissues, iridescent cellophane and bright ribbons wrapped around the handles.

Taste is another important consideration. For maximum freshness, small orders are advised for retailers. Lots of variety is a further attraction to customers.

One big challenge for retailers is the off-season slump. In the summertime, people lean toward cooler treats. "Don't let chocolate become a wallflower during the summer," warned chocolatier Lucille Hauser in an interview with *Gourmet Today.* "Encourage your customers to serve chocolate when entertaining. It is a perfect no bake dessert for the summertime." Chocolate gourmet sauces are also delicious on both ice cream and fruit.

CHIPS AU CHOCOLATE AND PUFFS AU PRALINE — Gift Packages

Valentine's Day gift packages of Chips Au Chocolate and Puffs Au Praline are wrapped with an abundance of ruffled pink cellophane. A pink ribbon laced through a heart-shaped cutout completes the holiday theme.

PUFFS AU PRALINE

A gourmet melange of Puffs Au Praline intimates a party atmosphere with renderings of pink and blue streamers. "Puffs," in large white lettering, is backed by a casually drawn brush stroke in blue.

STIX AU CHOCOLATE

Dipped in white or dark chocolate, these extravagant pretzel rods are pictured with a decorative pink ribbon on a pink background. The Stix Au Chocolate box comes in a vibrant purple with a marble-like grained design.

CHIPS AU CHOCOLATE

The Elite Treat offers a gourmet delight with chocolate-covered potato chips available in a stylish metal canister or an ice-cream type of container. The canister's lid shows a partial photograph of rich-looking chips in white and dark chocolate, while the brown cardboard container is covered with a gold lid. The logo of dripping chocolate appears on both.

LONG GROVE

Opulently decorated heart-shaped box places lace on a red-and-gold background. Motif continues with inset illustration of cherubs carrying hearts.

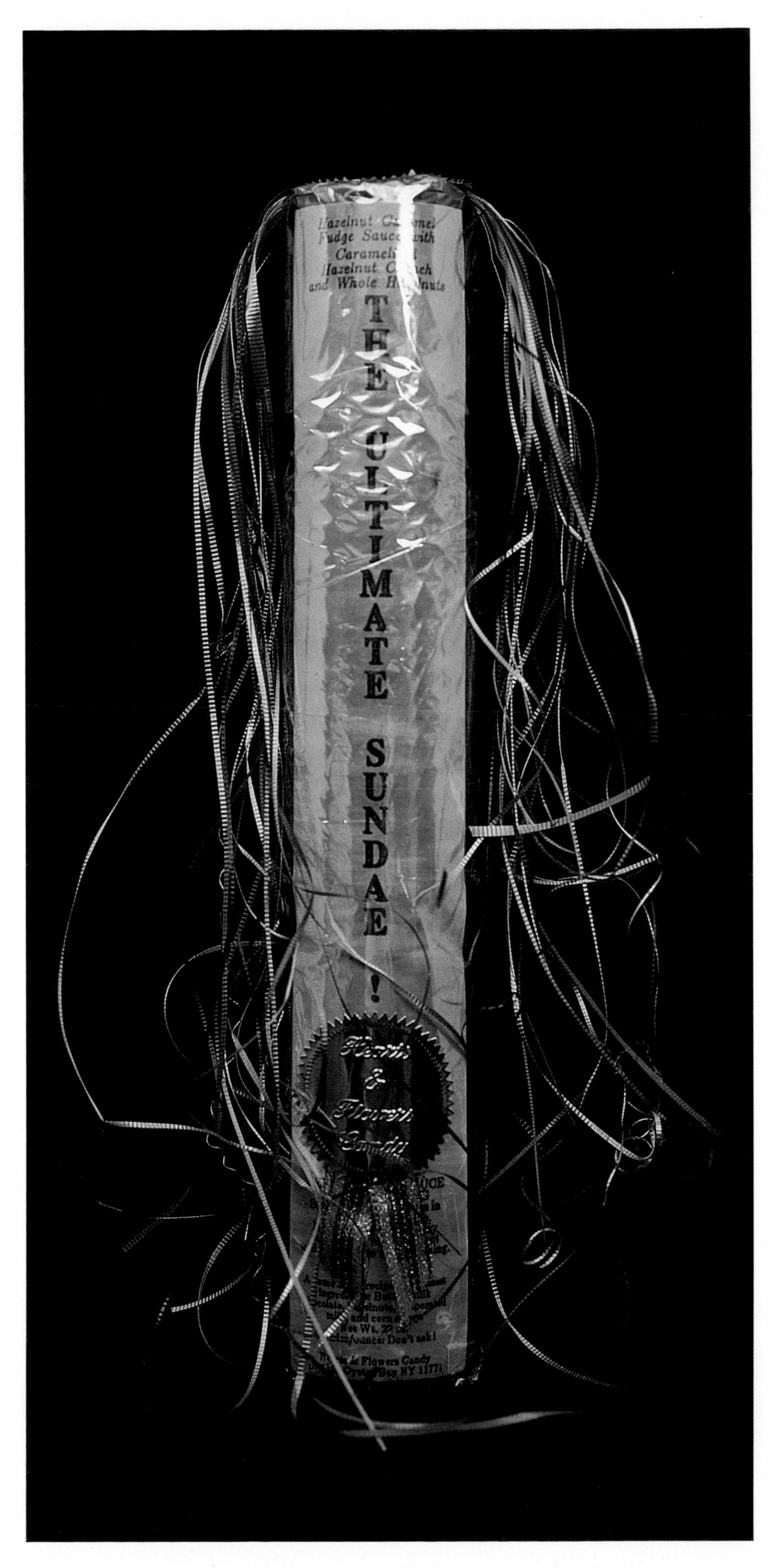

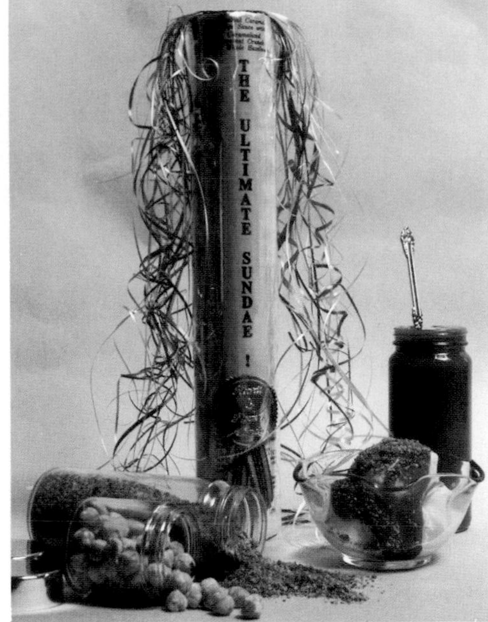

HEARTS & FLOWERS ULTIMATE SUNDAE

The Ultimate Sundae employs a party theme with multi-colored streamers and a gold-stamped blue ribbon. The individual jars of topping are wrapped in multi-colored cellophane.

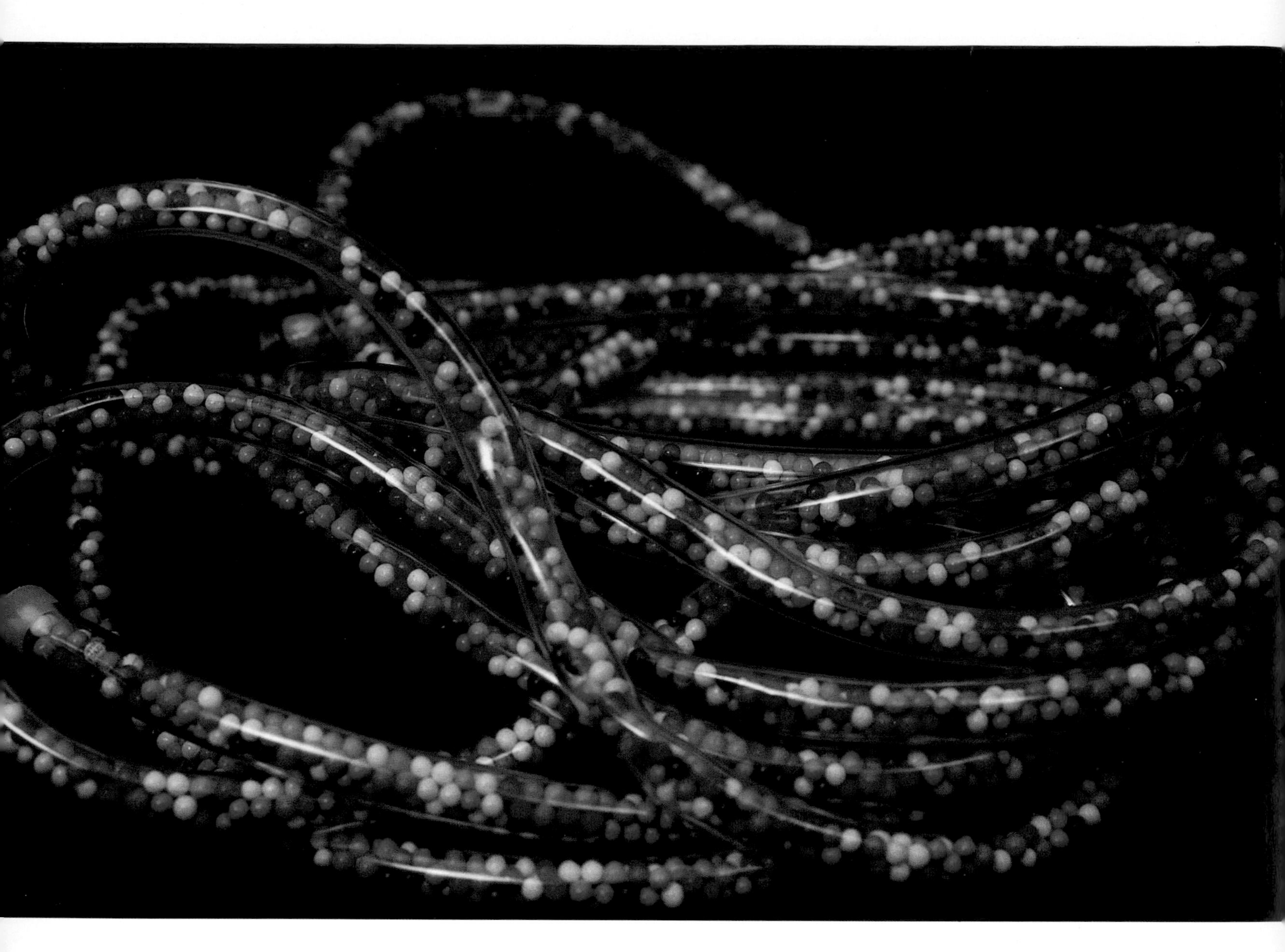

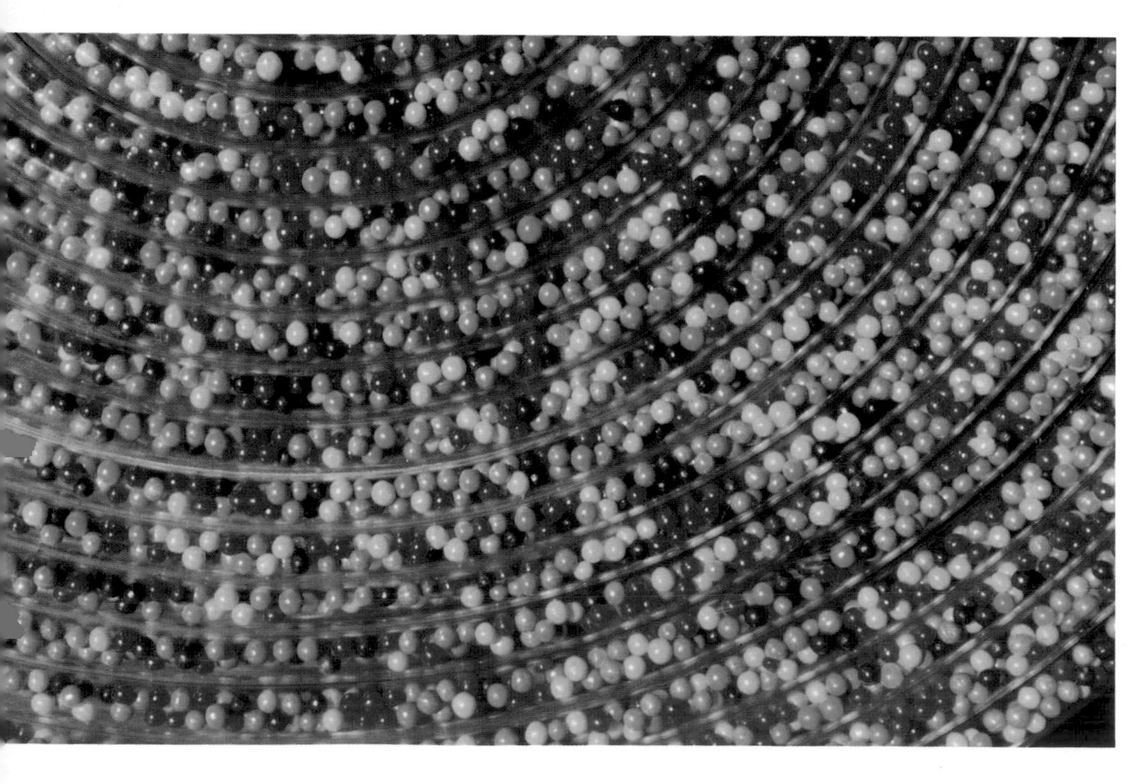

HEARTS AND FLOWERS RAINBOW ROPE

Hearts and Flowers' Rainbow Rope can be cut any length, up to an outrageous 100-feet of plastic tubing. This gourmet fruit-flavored candy combines innovative packaging with wildly bright colors.

ALOHA PASSIONS — Macadamia Nuts

Gold-embossed script on glossy-black paper portrays the chic, upscale image of Aloha Passions' macadamia nut candy. A gold elastic seal contributes further to the elegant presentation.

LONG GROVE — Chocolate Nutcracker

ASTOR'S CHOCOLATE SEA SHELLS

Embossed presentation of chocolate dessert shells highlights a blue box. Astor's distinguished emblem, in gold trim, appears in the upper corner.

MAGGIE LYON DOGWOOD SAMPLER
— Chocolates

A unique geometrically shaped box is beautified by a lovely dogwood-flower design. The finishing touch for Maggie Lyon's Dogwood Sampler is a paper dogflower attached to the top of the box, that serves as the closure.

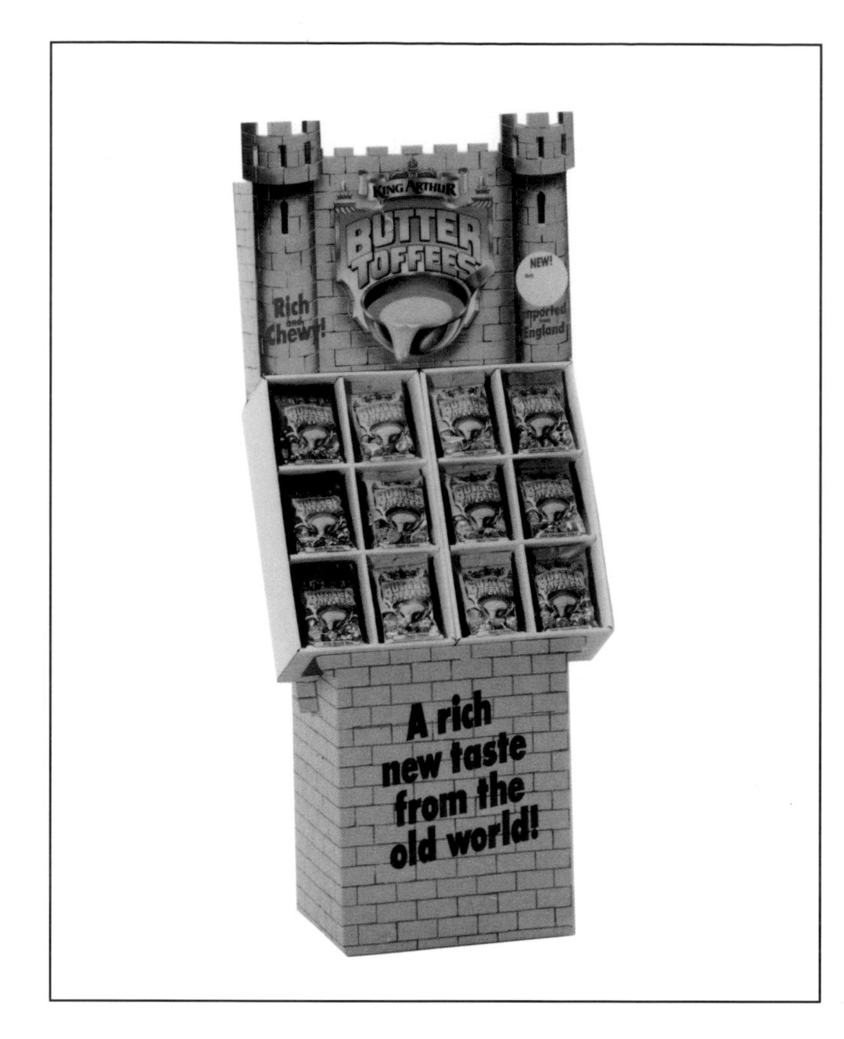

KING ARTHUR BUTTER TOFFEES — In Bags

A gold ladle is the focal point of King Arthur's Butter Toffees in red, yellow and orange bags. Like its counterpart in plastic containers, the label suggests an air of English royalty.

KING ARTHUR BUTTER TOFFEES

Butter Toffees are individually wrapped in gold foil and are partially visible from inside a plastic container. King Arthur's royal motif includes a king's crown and a coat-of-arms around a ladle.

LONG GROVE'S — Valentine Day Confectionery

Long Grove's off-white box is textured with an understated floral print. An illustrated red rose below a Valentine's Day message adds a splash of color.

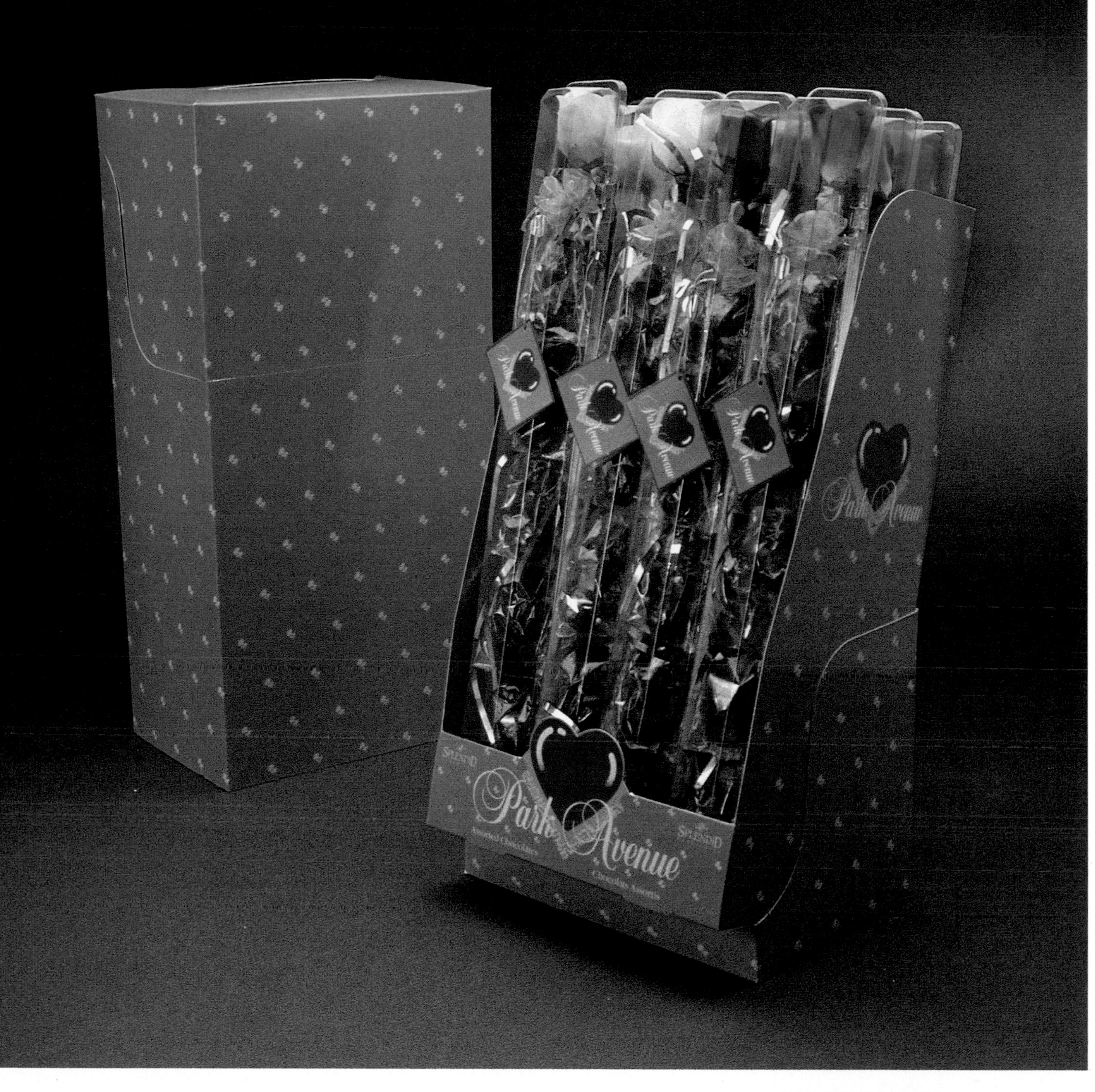

PARK AVENUE ROSES

A pink Valentine's Day display in an upright stance extolls thc cxtravagance of long, chocolate-laden roses. Park Avenue maintains the holiday's heart motif with renderings on product and on the front and sides of box. Roses are shown in pink, white and red.

Satin

CHOCOLATE MACADAMIA SATIN STICKS

Chocolate Macadamia Satin Sticks retain a metallic look with a polished metal cap and the glossy silver sticks which are visible through a clear jar. "Satin Sticks" are typeset in shadowed form for a three-dimensional effect.

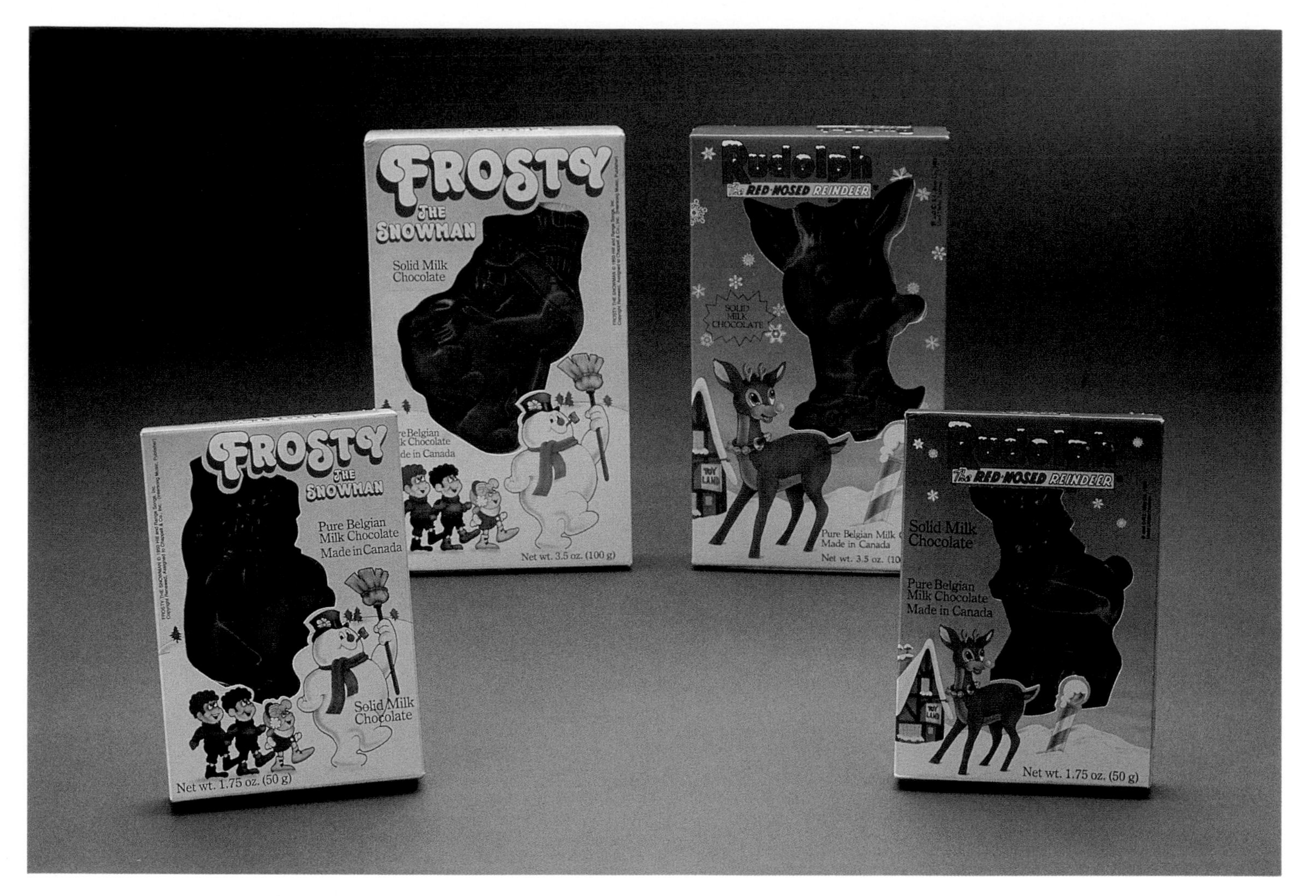

CARTOON FIGURE CHOCOLATES

Chocolate figures of Frosty the Snowman and Rudolph the Red-Nosed Reindeer are modeled directly from the popular Rankin-Bass cartoons of the 1960s. Each box features an animated scene reproduced from those television specials.

TALIAHAUS CHOCOLATES

Taliahaus' upright silver box infers sheer sophistication, further accentuated by a green, red and white bow. The regal Taliahaus emblem is printed in light, unobtrusive gold.

WILKINSON'S CANDY ALLSORTS

Wilkinson's Original English Candy Allsorts are aggressively designed for maximum appeal. A rainbow of colors in a variety of modern geometric shapes are shown in a photographed pile on a cardboard carrying case.

CLARIDGE'S TRADITIONAL ASSORTED CHOCOLATES

A black velvet ribbon on a dark box with fancy print signifies the gourmet appeal of Claridge's Chocolates. A horse-and-carriage silhouette in gold emphasizes the company's "traditional" heritage.

CLARIDGE'S TRADITIONAL ASSORTED CHOCOLATES

Claridge switches to a pink, red and green floral design for more of their Traditional Assorted Chocolates. The horse-and-carriage logo is carried over from their black box.

ASTOR — A Chocolate Party

Several concurrent graphic elements share the box, as would invitees of "A Chocolate Party." Four photographs of upscale images (servant, silver trays, etc.) line a gold box subtly imprinted with figures of affluent guests.

ASTOR CHOCOLATE LIQUEUR CUPS

The astor box, in gold typeface and trim, features a photograph of a decidedly grandeur nature. Beneath a pouring bottle of Kahlua, eight chocolate liqueur cups in gold foil grace an elegant gold serving tray.

LONG GROVE CHOCOLATE WINE BOTTLE

Chocolate in a wine-shaped bottle has a wine-like wrapper around the "cork." The bottle sits in a canister which displays an evocative, old-fashioned illustration.

ISLAND PRINCESS

Black string is tied around a green wrapper for Island Princess' Macadamia Popcorn Crunch. The sand-colored label carries the native palm branches.

ISLAND PRINCESS — Macadamia Nuts and Tropical Treasures

A horizontal spread of fruit, displayed at the bottom of the carton, allows Island Princess to switch visual gears—the blue-and-silver combination works well in pursuit of a "tropical" aura.

Hand-dipped Macadamia Nuts hold to the standard company logo.

ISLAND PRINCESS
A Gift from Maui
MACADAMIA BRITTLE
ISLAND PRINCESS
A Gift from Maui
MACADAMIA POPCORN CRUNCH
NET WT. 26 OZS.
ISLAND PRINCESS
A Gift from Maui
MACADAMIA BRITTLE

BROWN & HALEY HARD CANDY

Colorful, warm-hearted illustrations embellish the lids of Brown & Haley's Hard Candy. Tropical Fruit, for example, portrays a luscious tropical scene of fruit and palm trees at a beach. The three different flavors each have corresponding drawings.

BROWN & HALEY CREMES

A unified color scheme dominates an otherwise dark pacakage of Brown & Haley's three different cremes. Red, aqua green and orange hues are found in company logo, name of flavor and wonderfully appetizing photograph of the product.

BROWN & HALEY'S — Macadamia Nuts, Signature Collection

Wholes and halves of Chocolate Covered Macadamia Nuts are pictured with flowers on a black box. Reflective floral designs garnish the perimeter of the box.

Brown & Haley's Signature Collection is an offering of their finest chocolates in a gold-trimmed package. An opulent table setting places fancy chocolates on a gold-trimmed plate next to an alluring flower arrangement.

BROWN & HALEY ALMOND ROCA

The box and tin both utilize a romantic red, pink and gold scheme for Valentine-season package. The contents are individually wrapped in fancy gold foil.

BEVERLY HILLS CONFECTION COLLECTION

The Beverly Hills Confection Collection illustrates a distinct likeness to the city for which it's named. Sophisticated pink boxes, some tied in gold elastic, demonstrate palm trees and an affluent home boasting California architecture.

KRUMS
CHOCOLATIERS SINCE 1901
We're makin' it just for you
NET WT. 1 OZ.
KRUMS
We're makin' it just for you
Happy
Hanukah
HAPPY
HANUKKAH
KRUM'S
CHOCOLATIERS SINCE 1901
Coconut Bar
KRUM'S
Milk Chocolate with Almonds
NET WT. 2½ OZ.

KRUM'S CHOCOLATES

Chocolates for the Jewish New Year are contained in boxes with carefully arranged "hills" of Krum's ingredients.

Krum's Chanukah Chocolates features "Star of David" pops, miniature "Macabees" and chocolate "draydels." The confections are all certified kosher and are packaged in "fun" figures and designs that are appealing to children.

GHIRARDELLI DARK CHOCOLATE SQUARES

A penny-style jar of high-impact plastic serves as a display case for dark chocolate squares in individually wrapped gold foil. Ghirardelli's eagle logo proclaims "premium quality" on a flowing banner.

TROMBLY'S CHOCOLATES

The lower left corner of each black box supplies an appetizingly descriptive text relevant to each particular version of Trombly's Chocolates. Photographs of the confections are arranged in an organized, angled pattern.

GHIRARDELLI MILK CHOCOLATE

Ghirardelli's chocolate-colored label envelopes a gold foil wrapper. The one-pound serving size is a "weighty" testimony to confectionary indulgence.

RICHARD H. DONNELLY FINE CHOCOLATE

Donnelly's aristocratic, geometric logo is in a royal color scheme of gold, black and red on a dark, textural box. "Richard H. Donnelly," at bottom, takes advantage of extra letterspacing and light, extremely condensed typeface for a distinguished effect.

LOU RETTA'S CHOCO-LINIS

Lou Retta's tortellini "look-alike" uniquely combines white chocolate with a hint of cheese flavor. Red and green stripes on a black box reflect the Italian origins of tortellini. The Choco-linis are visible through clear cellophane.

LAMME'S CHOC'ADILLOS

The top portion of the box combines large white type onto a photograph of Choc'adillos. The bottom half design is a reverse of the top, with the same photograph appearing within block typeface on a white background.

VARDA CHOCOLATIERS

Varda represents their chocolates with a stylized geometric illustration of "V" for Varda and an ultra-modern black-and-white color scheme.

DEARBORN'S — Chocolate Truffle Kit

Eyecatching colors and "floating" truffles are the primary elements of Dearborn's Truffle Kit.

DEARBORN'S — Chocolate Hazelnuts and Dessert Caviar

A product of international renown is projected with Dearborn's world map, listing localities such as "Baked Alaska."

The tradition of hand-picking hazelnuts is re-created at the top and bottom of Chocolate Hazelnuts' vertical canister.

TRUFFELINOS

Truffelinos' square box is tied with a delicate pink ribbon. Below the gold Candy Jar logo, "Truffelinos" is splashed across the cover in oversized, ornate script.

MRS. BRITT'S CHOCOLATE HAZELNUT CRUNCH

Mrs. Britt's Chocolate Hazelnut Crunch makes its statement with a genuine Oregon wood lid and a tied bow of gold elastic. The opulent presentation continues with flowing script on a gold-trimmed label.

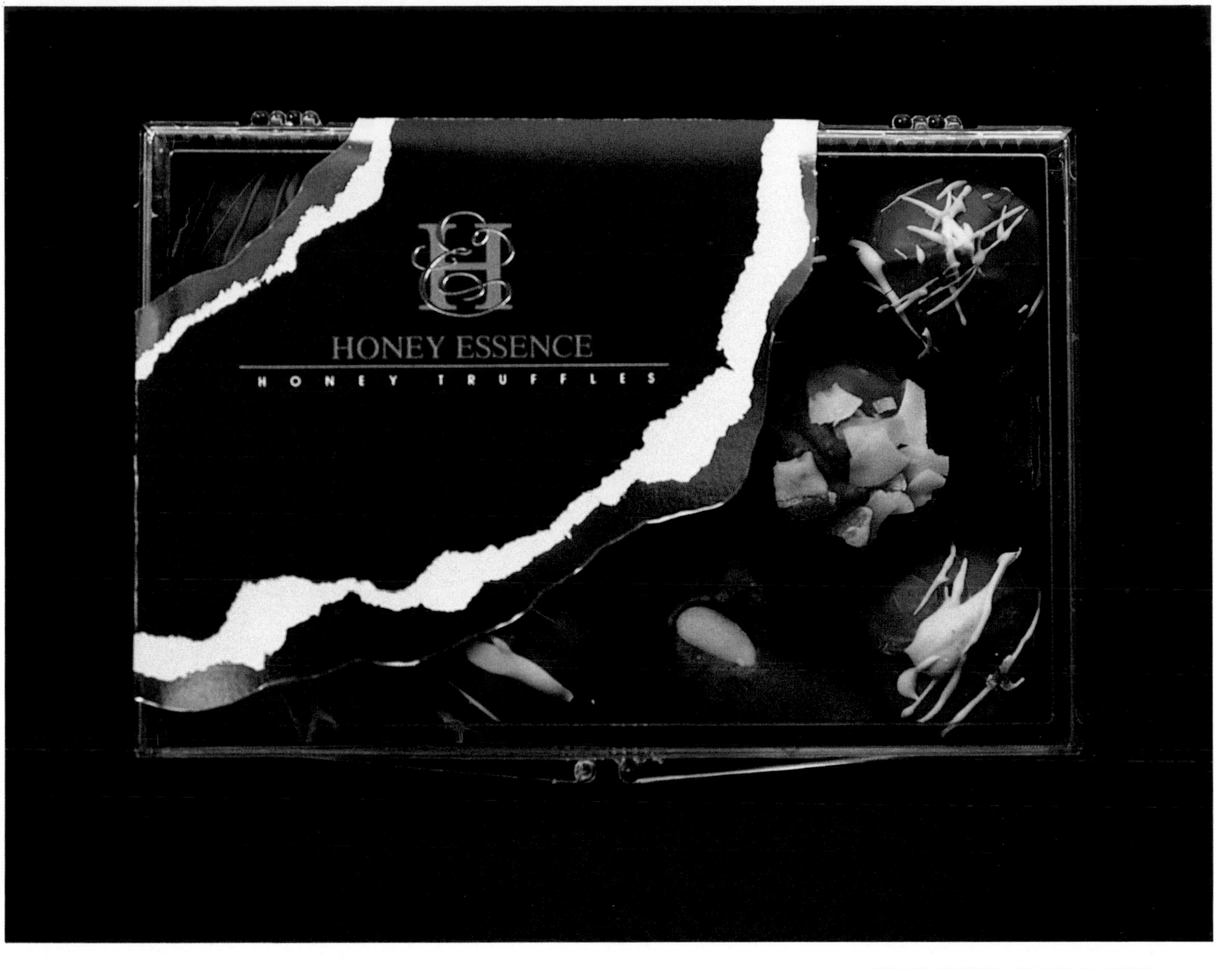

HONEY ESSENCE HONEY TRUFFLES

The Honey Essence label replicates a torn piece of paper for a frenzied, unpredictable look. The clear plastic box exposes several of the exotically created Honey Truffles.

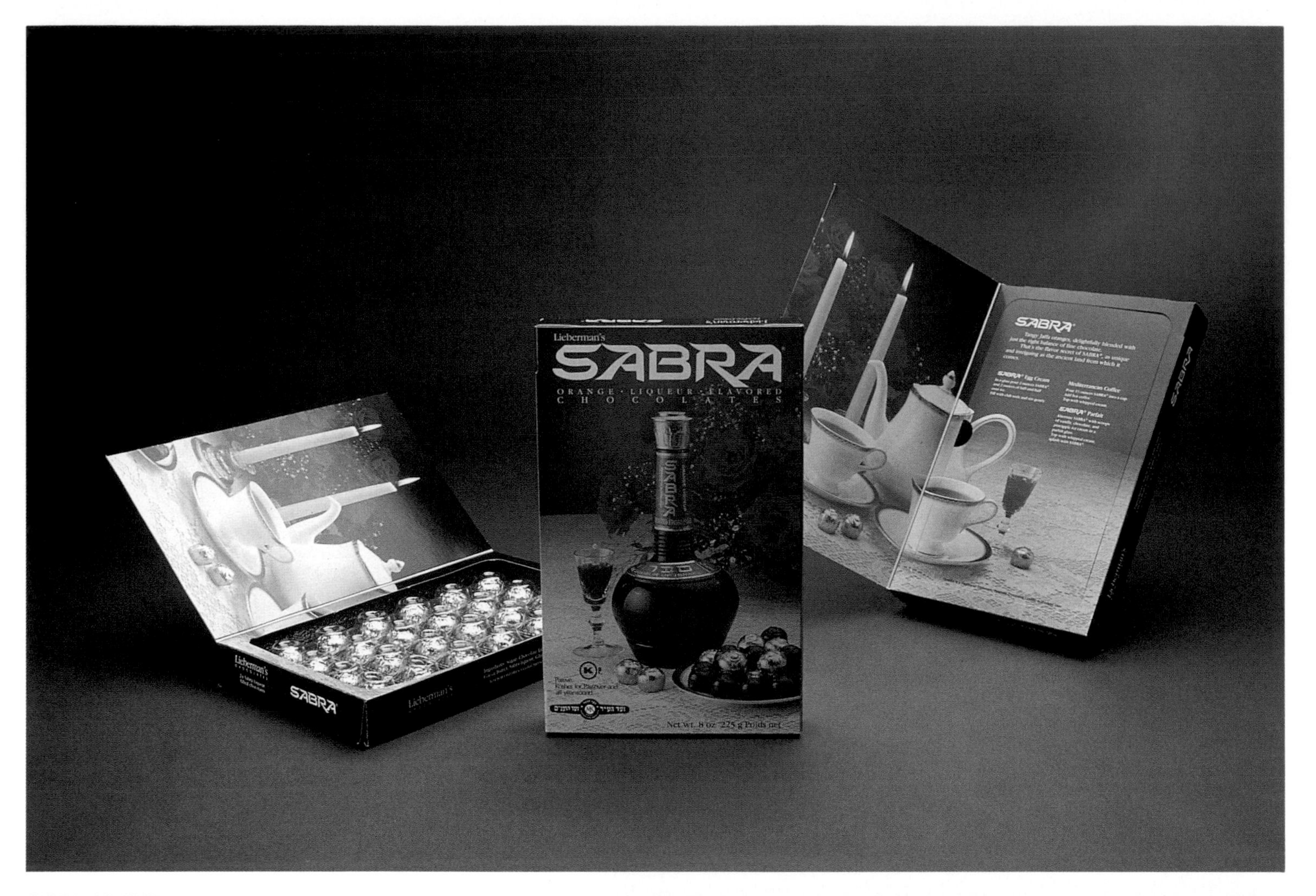

SABRA ORANGE LIQUEUR FLAVORED CHOCOLATE

Sabra Chocolates spotlight their flavoring with a cover photograph of an orange liqueur bottle next to roses and the product itself. Wrapped in gold foil, the chocolates are housed in a hinged box which opens to reveal an elegant table setting.

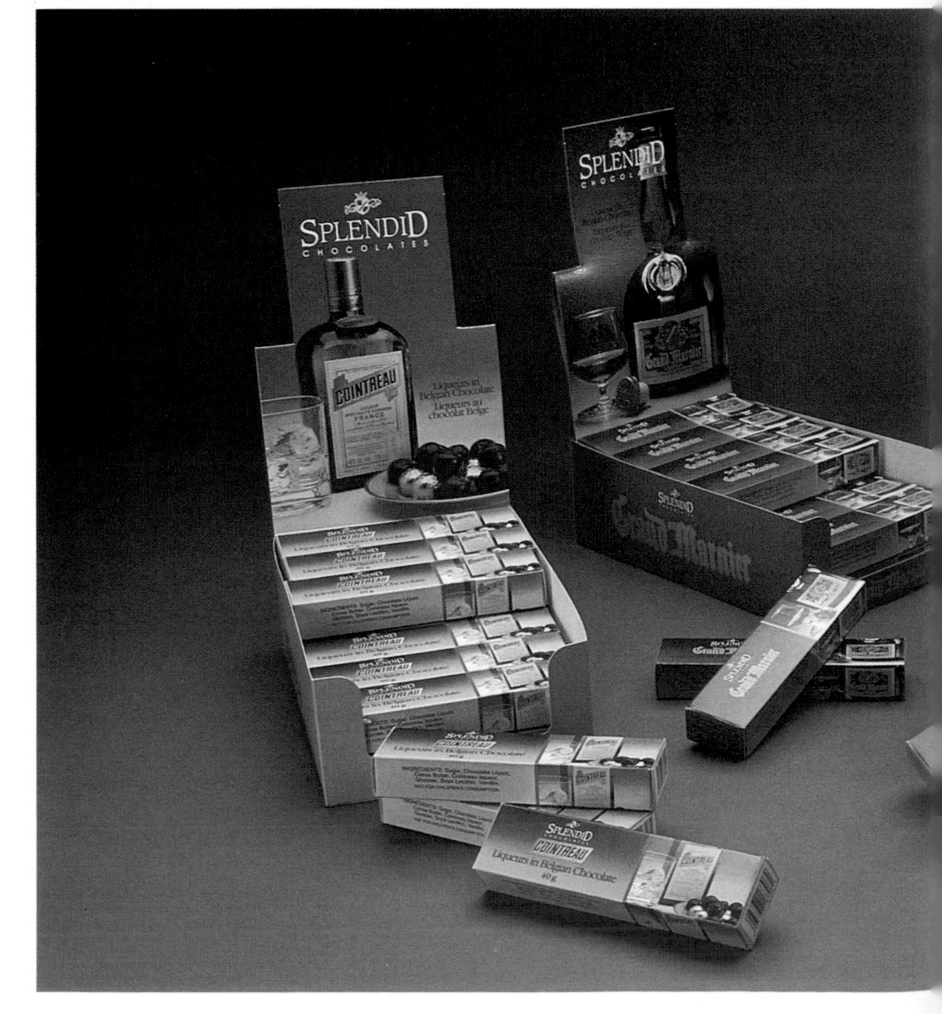

VAN DOUGLAS CHOCOLATES

Black, silver-trimmed boxes lend a simple but sophisticated appearance to Van Douglas' Chocolates. "Amaretto," "Mocha," and "Cappucino" are presented in reflective lettering, the latter two in ornamented gold script.

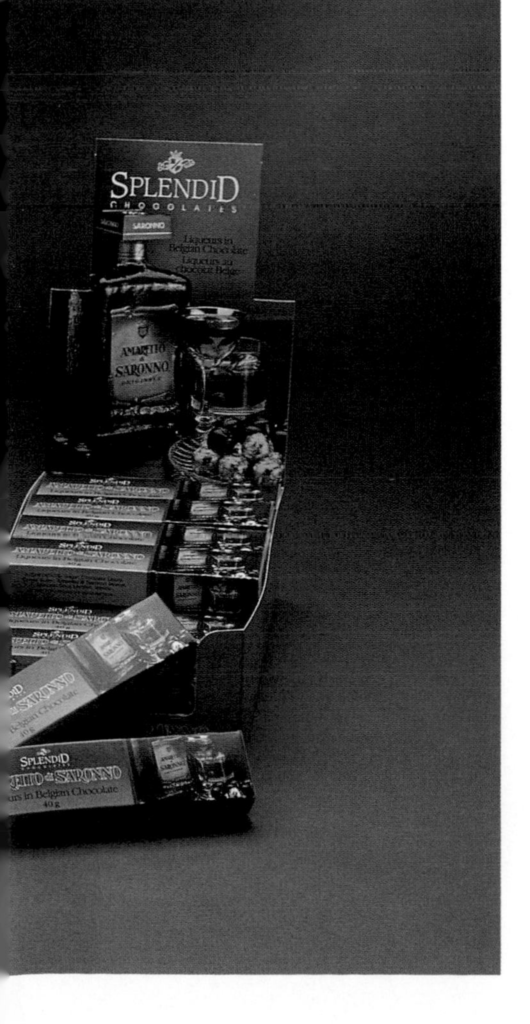

SPLENDID CHOCOLATE DISPLAYS

Three displays for chocolate liqueurs each highlight the product and a bottle of its derivative liqueur. The Cointreau display has a silver color scheme, the Grand Marnier an orange scheme and the Amaretti di Saronno an ice-blue one.

Long-Stemmed
CHOCOLATE
Roses
from
chocolate Whimsey

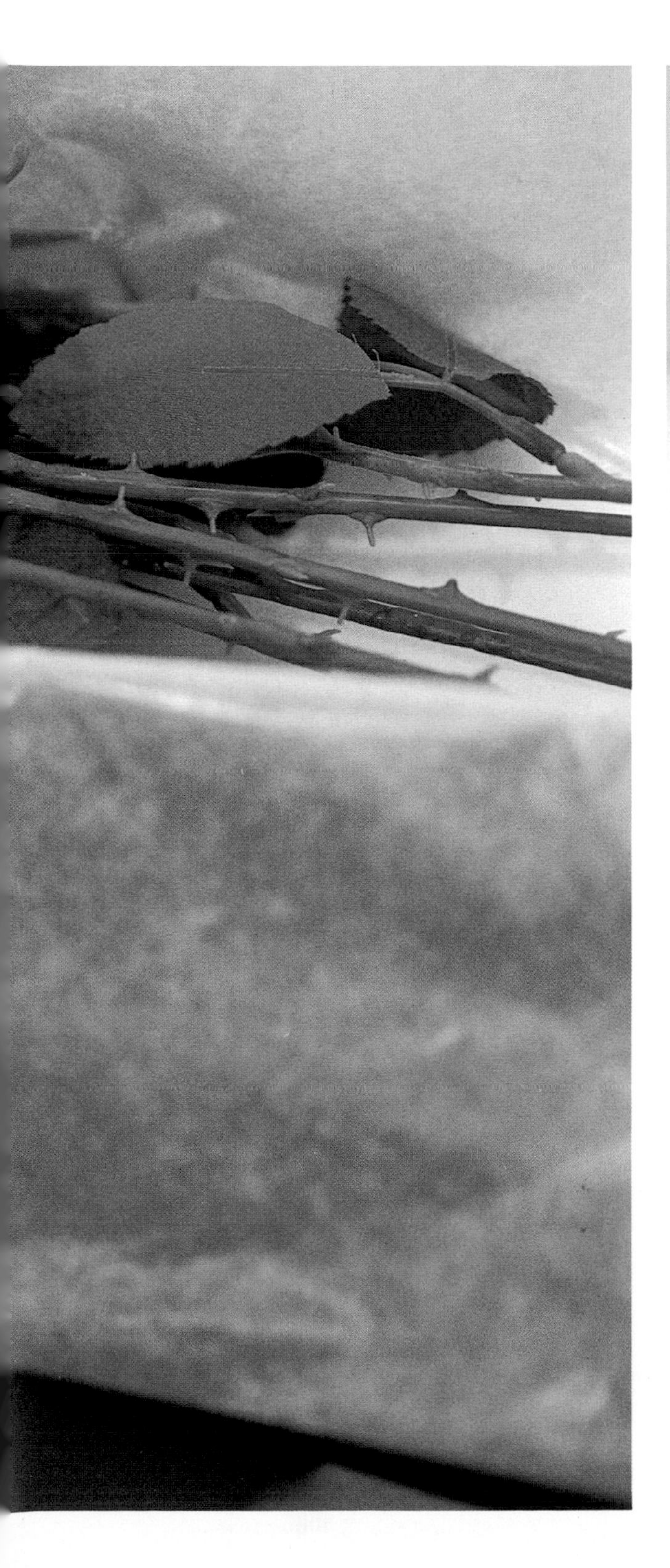

CHOCOLATE WHIMSEY — Long-stemmed Roses

A brown florist's box, resplendent with a brown satin ribbon and a beautiful gold label, is a perfect accompaniment to the dozen long-stemmed chocolate roses.

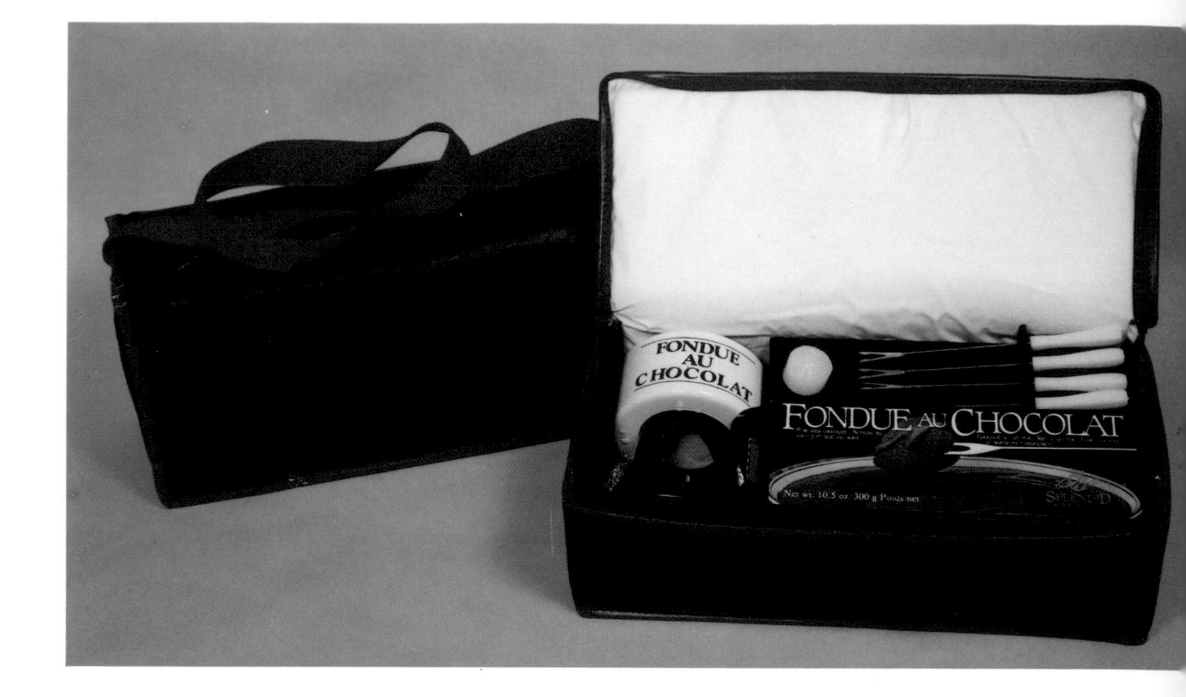

BOUNDARY WATERS YARD OF BUBBLEGUM

The Boundary Waters label depicts a yellow yardstick reminiscent of elementary school classrooms. The Yard (literally) of Bubblegum, in a clear plastic tube, is a playful item that kids will enjoy.

SPLENDID CHOCOLATES
FONDUE AU CHOCOLATE

A long horizontal package accents a delectable bowl of Fondue Au Chocolate. Slices of an orange and a strawberry are speared by a silver fondue fork and tantalizingly dipped in chocolate.

Splendid Chocolates' Fondue Gift Pack provides white fondue forks and a fondue warmer in a black-and-red carrying case.

Chapter Two

CRACKERS
PASTRIES
BISCUITS
COOKIES
CAKES

This may be the age of carrot juice and calorie counting, but baked goods still reign as the ultimate convenience food.

Such convenience originates from the gleaming glass-and-brass counters of commercial bakeries, as busier lifestyles and the quick gratification of impulse purchases have created a booming market for gourmet baked goods.

When customers walk into a bakery boutique, they look at the dessert counter and say, "Mom used to make these." Retailers such as a cookie shop owner in River Forest, Ill., report successful sales of homemade sweets. "We supply cookies to Grandma when her grandchildren come to visit. These days, Grandma is too busy to stay home and cook."

"Who would have thought a dozen years ago that people would pay so much for cookies?" asked futurist and author John Naisbitt in a forum sponsored by *Milling and Baking News*. Bakery products were a $22.2 billion industry in 1986, according to the U.S. Department of Commerce. Wholesale baked goods dominate the industry but retail and in-store bakeries both are showing healthy growth.

Consumer tastes, however, do shift. The croissant fad is on the wane and in recent years, dozens of shops in the cookie or muffin niche have been absorbed by national chains. As a result, one-item shops are expanding their offerings and jumping on the specialty bandwagon. Some niche stores are even adding soups and sandwiches to their pastry lines.

Supermarket bakeries have added new products, too. Muffins are particularly popular, while frozen cinnamon rolls are being offered by wholesalers for in-store baking. "More and more people," market analysts say, "are shopping for baked goods in a bakery under a supermarket's roof." In a 1987 trade journal survey, *Baking Industry* found that 90 percent of large supermarket chains and 50 percent of all supermarket companies planned to add at least one in-store bakery.

Image is another reason for the growth of baked foods. High-style merchandising is a distinct change from when full-line neighborhood bakers dominated the market 20 years ago. "Selling on image is new to the industry," said Peggy Hoffman, communications director for the Retail Bakers of America, in a 1987 interview with *Fancy Food*. "There's much more competition for the bakery food dollar than we have ever seen before. When (entrepreneurs) open up a store, they really market the product. They open up awareness."

With quality ingredients and an elegant appeal, specialty baked goods have taken over 40 percent of the market. Because special treatment brings higher sales, cake retailers are emphasizing presentation and appearance. "The more highly decorated a cake is, the higher the markup," Hoffman explained. "Take notice of fresh toppings and fresh fruits, even fresh flowers, on cakes."

One gourmet bakery in Encino, California generates half of its sales from cakes packaged in gift baskets. The cakes come in varieties such as Banana Pecan, Apple Walnut and Macadamia Nut, and can be stocked with any number of delicacies.

A new image was the winning formula for a Cleveland-based bakery. From a traditional small-bakery business offering cheesecakes, the company opened outlets with a turn-of-the-century Viennese look, featuring mahogany and brass cases in a 32-seat store. The patisserie now sells an assortment of flavors ranging from a standard carrot cake topped with handmade marzipan, to a chocolate mousse blackout cake.

Cookies, crackers, cakes and pastries are also being shipped around the country. Is a frozen cake, sent via Federal Express, worth the price? Yes, says a retailer who sells more than 10,000 cakes a month to homes and offices. Insurance companies send goods to policy holders, banks to new account holders, and airlines to frequent flyers.

Biscuit makers have also joined the low-sugar, all-natural trend. Today's biscuit lines feature all-natural oatcakes which are free from sugar, artificial coloring, flavoring and preservatives.

Cookies, meanwhile, have not lost their appeal to the American palate. The chocolate chip cookie found its way into American culture six decades ago, when Ruth Wakefield of the Toll House Inn in Whitman, Massachusetts, stirred pieces of bittersweet chocolate into a butter cookie recipe. According to the Retail Bakers of America, we gobble up more than one and a half billion treats from specialty cookie stores each year.

Price sensitivity varies from market to market. A retailer who caters to gift shoppers by offering collectible tins of cookies or crackers may cut out the customer who wants something special for the dinner table. Fortunately, the development of new packaging has answered the consumer's economic demand for affordable, practical items.

Not only do today's baked goods meet many market needs, the products invoke a sense of tradition while offering the whimsy and playfulness of everyday treats.

Nowhere is the nutritional value of foods more apparent than in the arena of breads, grains and pastries. The smart buyer has no reason to worry about chemical additives, preservatives or herbicides. More breads are made with whole or unprocessed grains, seeds and sea salt, and most packages boast fiber content or natural texture. Untouched by preservatives, the breads are crunchy, textury, earthy and wholesome.

Specialty shops are promoting Armenian Thin Bread, Stoned Wheat Wafers or "Par Baked" (partially baked) French breads. One store in Bay Harbor Island, Fla., uses only natural ingredients such as unprocessed grains, seeds, yeast, water and sea salt. Its products contain no fats, no oils, no preservatives and no sugar—a wholesome addition to any diet.

Another store offers beer bread mixes, in which the customer "simply adds beer and bakes" for a perfect complement to soups, salads or spreads. For sophisticated bread lovers, some shops offer cocktail toasts, a collection of hors d'oeuvres, or snacks and toasts with long shelf lives.

No food lends itself more to variety and improvisation than pasta. Pasta is, by definition, an Italian product. But mention pasta to an American and most likely his first thought would be of spaghetti or macaroni. To Italians, however, pasta embraces virtually an infinite range of varieties. Not surprisingly, pasta outsells all foods of Italian origin in the United States. Each American now consumes an estimated 12 pounds of pasta per year—far from the Italians' 55 pounds annual per capita consumption.

There is now an increased emphasis on specialty shapes, short cuts and tricolor pastas. A new line of domestically-produced pastas is available in fettucine, linguine and spaghetti, all sold in several flavors. Colored, flavored and filled varieties have quickly caught the public's fancy: pumpkin, squid ink and chocolate, among others.

Rice is another complex carbohydrate gaining favor among discriminating gourmets. Along with pasta and potatoes, rice is touted as one of the carbohydrates that are quickly and efficiently used by the body for energy. With only about 62 calories in a half-cup portion, rice fits into a reducing diet as well as one requiring well-balanced meals. Rice is low in sodium and has virtually no cholesterol and only traces of fat. Its protein content is excellent, nutritionists say, second only to soy beans among cereal grains.

But if the thought of rice once conjured images of a fluffy bowl of white rice served with stews and Chinese dishes, it now has a number of uses. The market now offers varieties such as floral-scented jasmine rice from Thailand, basmati from India and arborio from Italy.

Part of the credit goes to America's endless fascination with ethnic cooking. Cajun cooking has taken the country by storm and there's not a jambalaya, gumbo or etouffee worth its salt that doesn't include rice among the ingredients. Many specialty food companies offer an authentic taste of New Orleans in packaged flavored rices, with just the right blend of those ragin' Cajun spices.

Wild rice, the "caviar of grains," is a favorite of epicureans. An aquatic grass cherished for its nutty taste and tender, chewy texture, wild rice is often packaged with white or brown rice or other complementary grains. Several companies have recently introduced microwaveable wild rice that cooks in 10 minutes.

The dull staples, legumes and beans, are also commanding customer attention. In the past, legumes such as kidney or garbanzo beans, lentils, peas or Navy beans, were available in grocery stores and used as ingredients in soups or salads. No longer. Legumes are attracting more customers for their health benefits. "Beans are nature's most complete vegetable protein," said Jim Melban of the California Dry Bean Advisory Board. "There are very few foods today that can lower one's cholesterol (like) beans." When combined with cheese, corn, wheat or small amounts of meat, beans form a complete protein.

GRANDMA'S CAKES — Amaretto and Fruit

The logo on the right side of the box is a romanticized rural cottage typical of the "Grandma" who makes fruit cakes.

JAKE'S TRUFFLE CAKE

Jake's delectable Truffle Cake is luxuriously packaged in a unique octagonal box and illustrated with a photograph of the product on a lace placemat. The white label with seemingly floating script gives a light, dreamy presentation.

KEEP REFRIGERATED
JAKE'S
SINCE 1892
READY TO SERVE
HOMEMADE
Chocolate
Truffle Cake
NET WT. 16 OZ. (1 LB)
KEEP REFRIGERATED
JAKE'S
SINCE 1892
HOMEMADE
Chocolate
Truffle Cake
NET WT. 4 OZ. (32 g)

SNOQUALMIE FALLS LODGE

The line of Snoqualmie Falls Lodge products possesses delightfully picturesque design elements. With drawings of a soothing waterfull, a rainbow and Christine & Rob's home-style items, an appealing natural look is achieved.

GLENNY'S RICEWICH

The organic nature of Ricewich is exemplified with earth-toned browns on the label and a rendering of harvested rice.

SORRENTI'S WILD RICES

A vivid yellow ribbon, large clear typeface and a simple white bag contribute to a refreshing directness of message. Each package of wild rice has a tastefully colored drawing of its ingredients.

BEST OF LUCK COOKIES

Best of Luck's namesake is superstitiously symbolized with horseshoe shapes which frame several smaller horseshoes. Inside the metal box, the cookies are shaped into horseshoes and horseshoe nails.

WALKER'S SHORTBREAD COOKIES

Majestic portraits of old European military scenes unite with a strong plaid background to evoke the Scottish influence of this shortbread.

GOLDEN WALNUT COOKIES

Sharply photographed walnuts stand apart warmly from the black box, as does "Golden Walnut Cookies" in orange type.

TABOULI — Salad Mix

An intricately arranged platter of lettuce, tomatoes, cucumbers and olives encircles the mosaic appearance of Tabouli Salad Mix.

NATURE'S BURGER

On a simple, off-white box, Fantastic Foods shows confident pride in their meatless burger mix with a large scale photograph of the item.

SAN ANSELMO'S BISCOTTI

This elegant Italian treat, in a bakery-style bag, conveys an opulent appearance. Pink roses, a gold coffee cup and a lace-style mat beneath the product suggest the same.

VICTORIAN TEA BISCUITS

The dainty decor and pink ribbons of Victorian Tea Biscuits harken to the atmosphere of English tea parties. The cellophane-wrapped almond cookies are arranged in neat rows and nestled on a decorative paper doily.

CHOCOLATE DECADENCE, CHOCOLATE & THUNDER WHITE LIGHTNING

Slices of each of the chocolate cakes are invitingly pictured on distinctive silver platters. Ornate, flowing script reflects the richness of the dessert.

EFFIE MARIE'S

A conservatively colored flower print projects an ambiance of good-old home cooking. Effie Marie's profile is flanked by text of her cooking philosophy and an old-style drawing of her country kitchen.

BRENT & SAM'S COOKIES

Chocolate-toned packaging is repeatedly used for these cookies. Caricatures of Brent and Sam visually indicate the "handmade" claim of the company. A clear cover and a gold elastic bow add distinction to the metal container.

SUSIE Q'S PINQUITO BEANS

The cheesecloth sack is a preferred packaging material for bean distributors. Susie Q's bold, artistic lettering highlights the company logo.

LITTLE BEAR REFRIED BEANS

A desert setting shows "Little Bear" wearing a sombrero, in earth tones which are consistent with the product's color.

EDEN BEANS

Four glass jars are labeled in corresponding color to the type of organically grown beans contained within. The rendering of two children in front of a crock of beans implies a product well-suited to family meals.

FRENCH MEADOW BREADS

Vibrant oval labels (in pink, powder blue and yellow) carry an appealing floral insignia around the perimeter.

LOAF AND KISSES — Croutons

A lace ribbon bedecked with romantic pink hearts decorate the oversized reusable jar. A family album photo and a nostalgic typeface also portray a homemade quality.

ESCOES'
ORIGINAL
Cheese Biscuits
16 OUNCES
•SECKINGER•LEE•

SECKINGER LEE'S BISCUITS — Escoes and GG's

Two metallic packages — a one-pound canister and a foil silver bag — are reflective and sleek, giving a streamlined presence to Seckinger Lee's biscuits. Widely-pinstriped labels augment the modern theme.

VERMONT FARE — Pancake Mix

Allusions to American heritage are evident in "Vermont Fare" 's Revolutionary War-era typeface and crowd of early settlers seated at a table under a tree.

DEARBORN'S — Flappy Jacks and What Waffles

Bright, tablecloth-like prints adorn Dearborn's novel triangular boxes. Modernistic renderings of pots, ovens and silverware are scattered about.

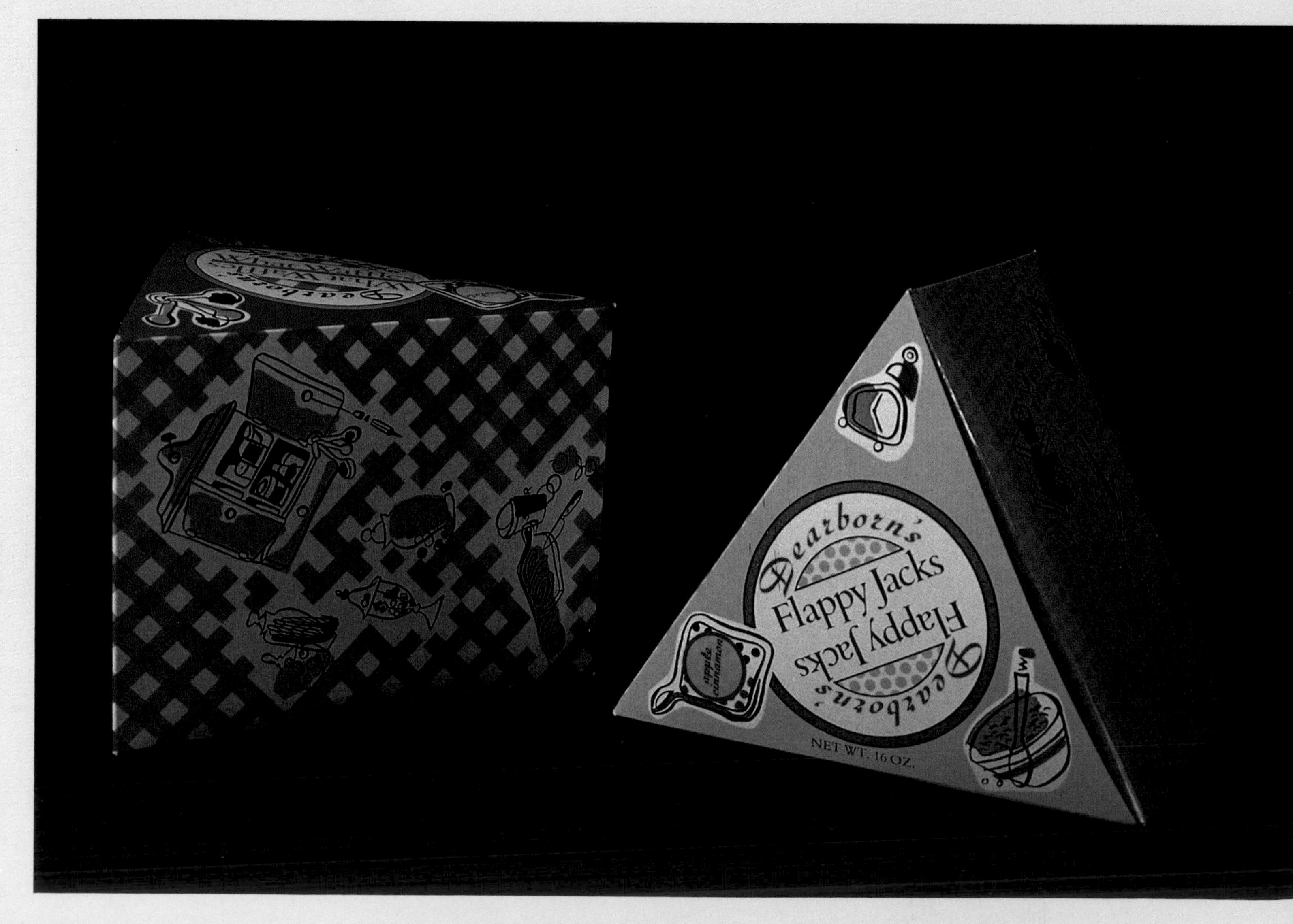

ANDRIZZI'S — Pancake, Crepe Mixes

A sketch of utensils and ingredients, on a kitchen wallpaper print, emphasize the satisfaction of creating fine foods at home.

JUST DELICIOUS

Pictured with a hand-woven cloth and a unique wicker creation, the Just Delicious Soup Mixes accent their homemade appeal.

ADRIENNE'S LAVOSH HAWAII FLATBREAD

Adrienne's Hawaii Flatbread employs a fitting Hawaiian motif with a tropical flower arrangement which serves as the Lavosh Hawaii emblem.

LE CROUTON

A white, bakery-type bag suggests that Le Crouton's will deliver bakery-type freshness.

MONICA'S COOKIES

A simple design of bright blue stripes is Monica's formula for attracting the buyer's eye to the goodies inside her cookie bags. The pecan and butter flavored cookies come as little balls wrapped in gold foil.

PUTNEY PASTAS

Transparent containers espouse the freshness of the various pastas. A photograph of a heart-rimmed plate provides convenient white space on which the ingredients and style of pasta are printed.

SAVOIARDI LADYFINGERS

A centered red banner leaves sufficient space to see Viero's wholesome Savoiardi Ladyfingers. A warm yellow color is taken by a drawing of eggs in a basket, the cellophane pouch and the product itself.

VIERO'S LE GOLOSERIE

Various shades of brown occupy Viero's clear packaging. Macaroons and "Ugly But Good" biscuits each display a silhouette of a mother and daughter sharing a dessert tray.

SELECT ORIGINS RICE BLENDS

Sample gift boxes of premium Texas Basmati, California wild rice blends and imported Italian Arborio rice use a porthole to show their exotic grains. Eye-catching script typefaces explain preparation methods and serving tips.

AUNT PATSY'S CHICKEN THYME SOUP

This package employs a transparent opening in front to entice the buyer with the colorful assortment of beans inside. A sketch of Aunt Patsy occupies the center portion of the logo.

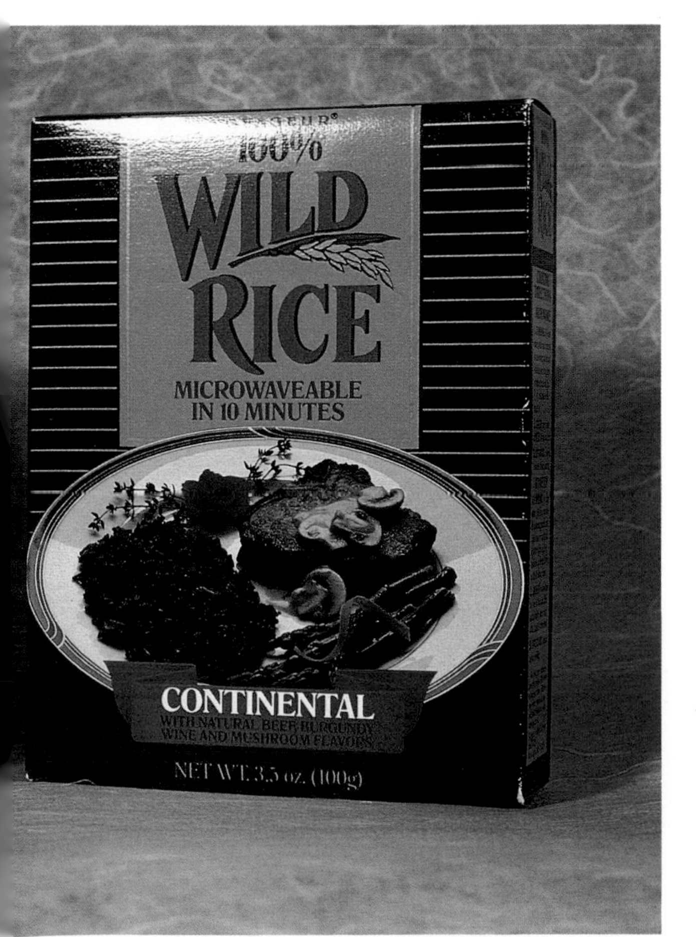

VOYAGEU R WILD RICE

The caviar of grains, wild rice is gaining ground quickly as an easy to cook item. Voyageur's uses bright typefaces to promote its wild rice. Photos of grilled salmon and beefsteak seasoned with spinach, herbs and other garnishings aim to show that wild rice complements all types of dishes.

HYE ROLLER SOFT CRACKER BREAD

A knife and cutting board on a kitchen counter are geared toward creative cooks who prefer unique cuisine. Cut and rolled soft cracker bread gives an unusual appearance and the predominance of brown suggests earthiness of the bread.

RUBSCHLAGER BREADS—Pumpernickel, Wheat, Rye

Tightly sealed clear plastic gives Rubschlager's Breads an appealing, just-out-of-the-oven look.

TAMARI BROWN RICE CRACKERS

The partially transparent pacalkge reveals the top halves of temptingly fresh-looking Brown Rice Crackers. Contrasting black, red and white copy with a Japanese inscription suggests authenticity.

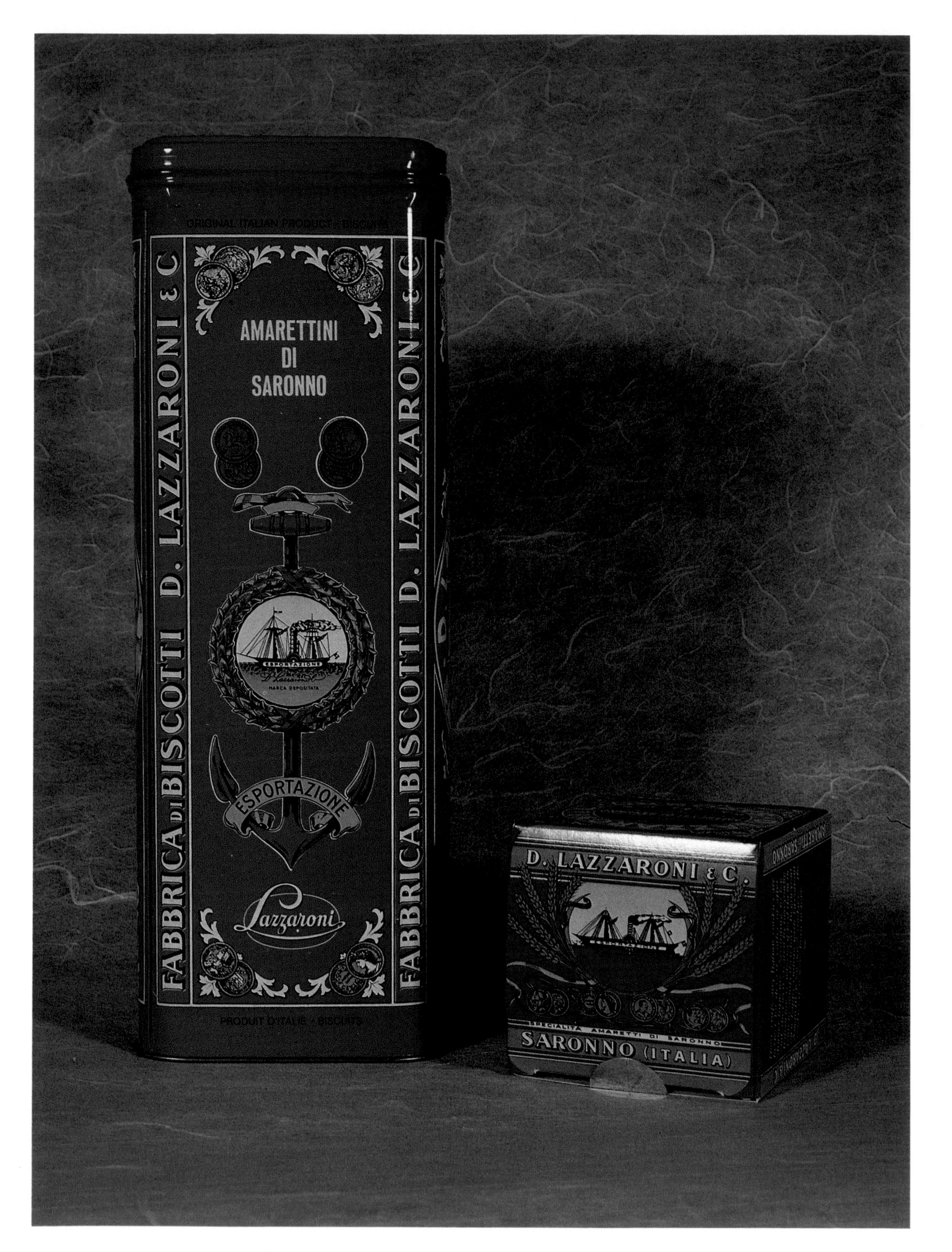
ORIGINAL ITALIAN PRODUCT · BISCUITS
AMARETTINI
DI
SARONNO
FABBRICA DI BISCOTTI D. LAZZARONI & C
ESPORTAZIONE
MARCA DEPOSITATA
Lazzaroni
PRODUIT D'ITALIE · BISCUITS
D. LAZZARONI & C.
SPECIALITÀ AMARETTI DI SARONNO
SARONNO (ITALIA)

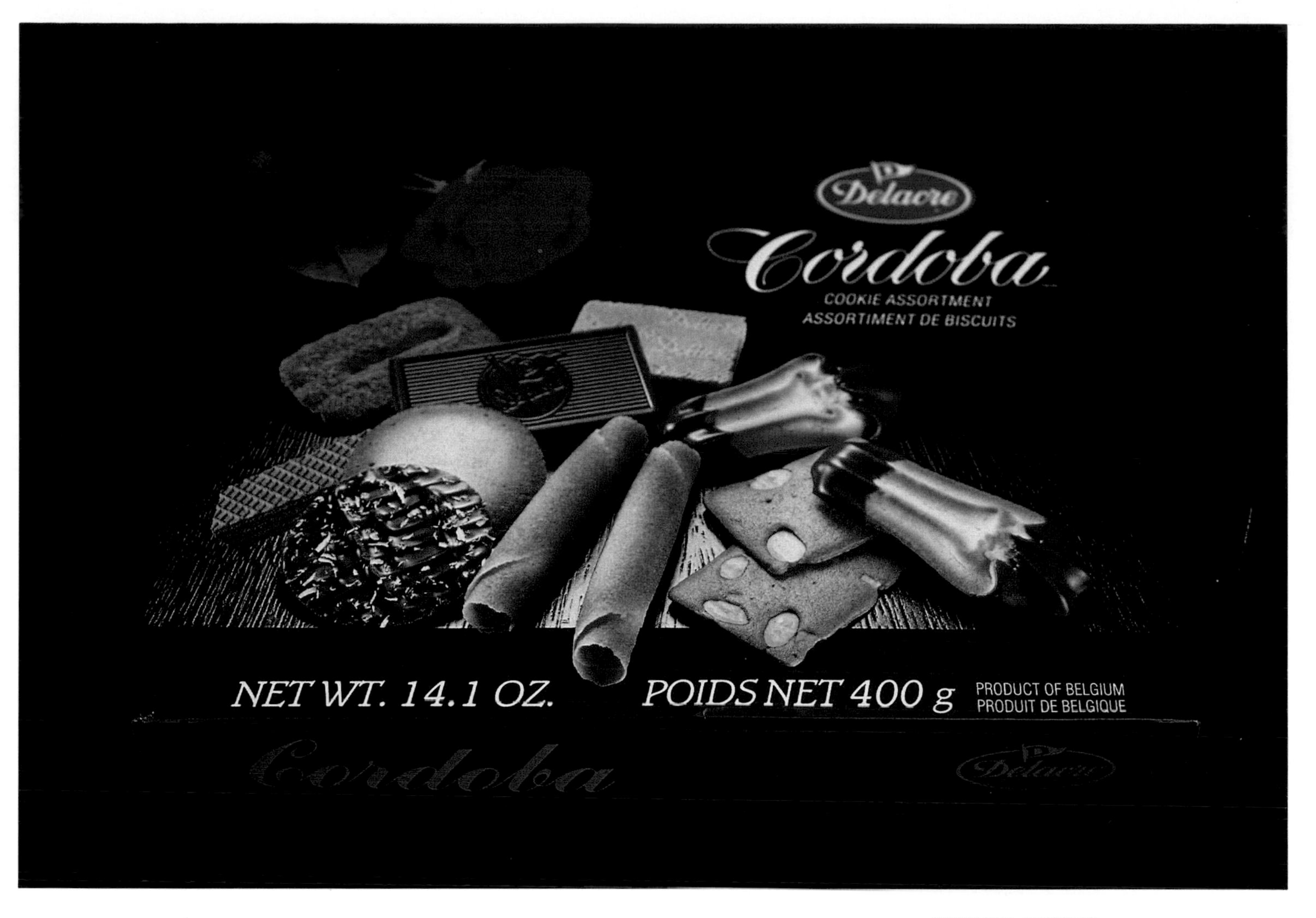

CORDOBA COOKIES

A beautiful arrangement of Belgian cookies is garnished with a red rose. Contrasting black and red colors bring out the texture of the cookies and a wood-grained table.

AMARETTINI DI SARONNO

A blaze of red and white colors adorns D. Lazzaroni and C. Saronno's collection tin. Packed in tall, rectangular and square cartons, this Italian import uses a ship and distinctive insignia to portray the product's age and reliability.

KOLLN CEREALS

Blue, red and brown are the primary colors of this simple package. A spoon in the bowl directs the eyes to an overhead view of the natural-looking product.

DASSANT — Bread Mixes

The face of the uniquely long, vertical bags carries a tastefully understated print design. Dassant's coat of arms adds a dignified flair.

MAYACAMA'S SAUCES

Close-up photographs of culinary delights dominate the face of the product. An ornate "M" headlines the package.

TRIOS PASTAS

Vividly clear packaging furnishes an appetizing view of Trio's linguini and tortellini. The pictured item signifies the primary ingredient.

FANTASTIC NOODLES

An overhead photograph of the whole wheat noodles is divided between the front and top of the package. The spectrum-colored Fantastic Foods logo appears in the corners of the bright yellow label.

CARR'S MUESLI COOKIES

Using a simple parallel-lined print with a photograph of a single cookie, Carr's Cookies feature an airy, easy-to-read design.

COOKIES BY CAROLINE

Cookies by Caroline employs a bright red heart and a yellow label to promote its nut-studded chocolate and chunky pecan cookie mixes. Alongside the close-up photos of the cookie pieces are reminders that this gourmet package contains "no cholesterol."

LARRY MULKEY'S ORIGINAL
CROISSANT CRISPS

The simplicity of the white, bakery-style bag assures the freshness of these croissant crisps.

ARVEY'S SPIRIT CAKES

Splashes of red, yellow and green stand out on the reflective silver boxes. Silver elastic and a black band give the octagonal boxes a gift package look.

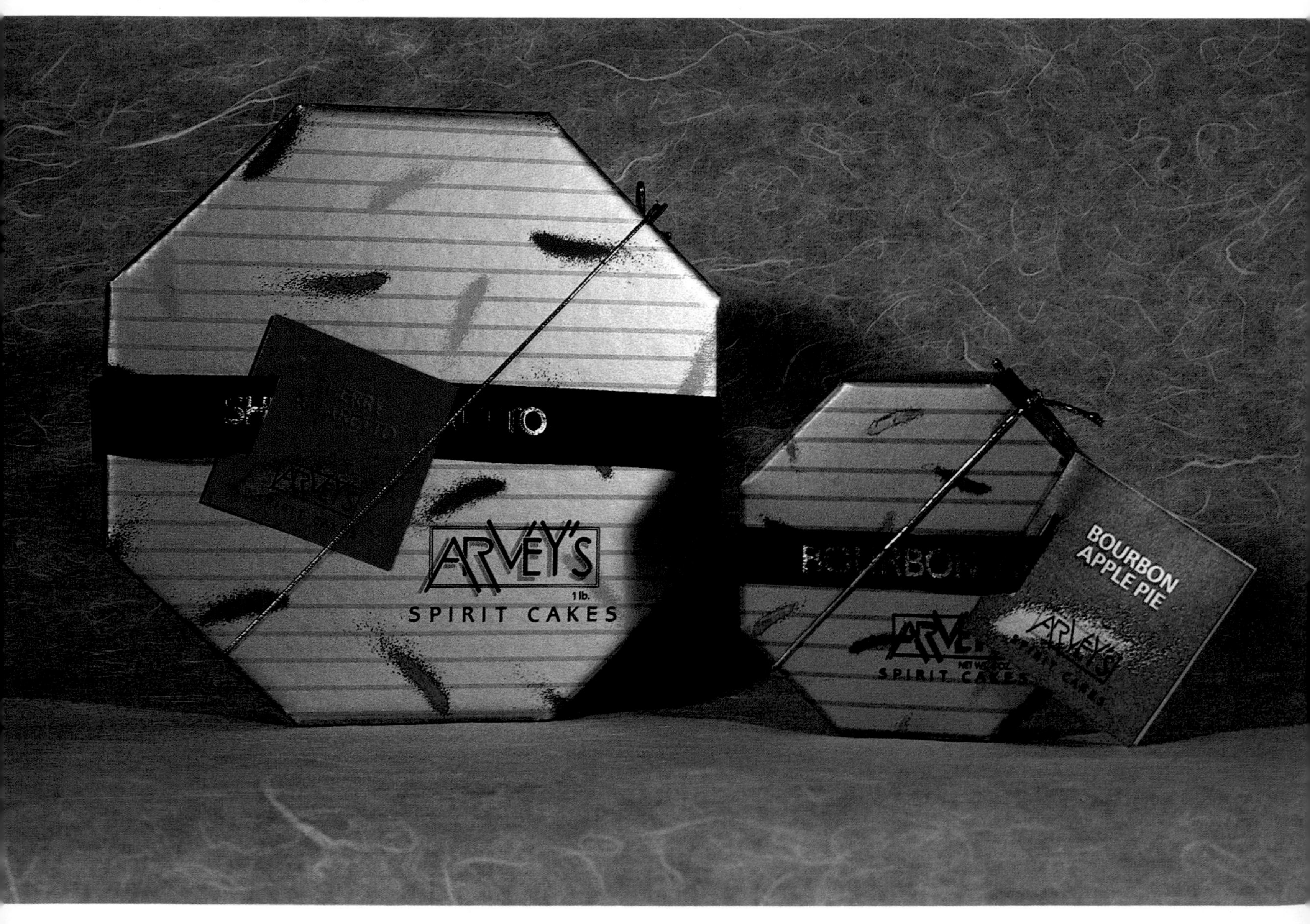

ROWENA'S ALMOND POUND CAKE

As seen through its clear plastic wrapping, this pound cake looks genuinely "wonderful" and freshly-baked. The bright red bow and a dainty flowered label confidently attest to merchandise which is homemade.

Chapter Three

SNACK FOODS
FRUITS
NUTS

What used to be known as snack food is today's "little meal." The half-sandwich, an appetizer and salad, a slice of pizza and bottled juice have taken the place of three-square meals a day.

Thanks to the microwave oven, better processing and creative packaging, time-pressed consumers can satisfy their demand for snack food. Market analysts attribute the rise of snack food sales to the fast-paced lifestyle of the '80s, the increase of women in the work force and the growing trend of Americans who prefer grazing or snacking rather than regular meals. Hot-selling snack-foods are potato chips, nuts, different kinds of cheeses, fried chicken, and salads. Despite the trends toward health foods, market studies show that high-fat ice creams, chocolates and triple cream cheeses appeal to the snack food crowd.

Intense competition among manufacturers has led to an abundance of new products, many of which are flavored versions of old favorites. Dried fruits have become a staple of many adult snackers' diets. Handsomely packaged as upscale gift and mail order items, dried fruits are in supermarket shelves as well as corner delis. Almonds still lead U.S. nut consumption in profitability.

Popcorn, the all-American snack, has grown 92 percent in the past five years. Americans now eat about 11 billion quarts annually or about 45 quarts per person, according to the *Lempert Report.* Nutritional authorities and leading health organizations such as the American Cancer Society and the American Dental Association have endorsed popcorn as a whole-grain, high-fiber, sugar-free snack.

A cup of air-popped popcorn without butter contains only 20 calories. Suppliers have responded by churning out flavors like Buttery, Natural Cheddar Cheese, Caramel, Macadamia-Honey, Nut, Pecan and Almond.

Though Americans profess an interest in foods that are low in salt and fats, they spent an estimated $3.3 billion on potato chips in 1986. Novelty ice cream products are crowding the half gallons in supermarkets and ice cream stores. Waffle cones and fresh peach, strawberry and chocolate brownie topppings are the rage.

Manufacturers are nonetheless aware that more and more customers are reading the labels and scrutinizing nutritional information. As a result, they are energetically creating low-fat, low-sodium and no-cholesterol foods. Their successfully redesigned products are now more upscale and convenient than ever.

EVE'S REVENGE

A dark, almost sinister motif begins with the emphatic underlining of Eve's "Revenge." Dark contents, a black label and a deep maroon ribbon combine on a sumptuous dipped apple.

BROWN & HALEY NUTS

Impeccably placed nuts line the perimeter of the Brown & Haley label. Nuts are partially illuminated — as are their reflections — for distinctive effect. Gold typeface symbolizes the copper kettle cooking process.

GOLDEN VALLEY PISTACHIOS

Pistachios in a wine carafe convey a gourmet appearance, while a crossed red ribbon over the neck touts the company's history. Pistachios in bag use the same, elegant red-and-gold label.

THE MAUI CHIP

An invitation to peer through palm trees reveals nearly every element of the 50th state's landscape. Two parrots and a native phrase in Hawaiian scroll adorn a spectacularly vibrant metal canister—a certain collector's item.

The same graphics, but with magenta trim, are used for the bagged version of The Maui Chip. A circular "stamp" indicates the "all-natural, no cholesterol" quality.

E.P. ANDREWS CONNOISSEUR'S POPCORN

These popping corns connote their "Connoisseur" character with a sophisticated, old-style portrait of E.P. Andrews and tasteful red lids.

ISLAND
PRINCESS
A Gift from Maui
HAWAIIAN PEARLS®
CHOCOLATE
COVERED
HAWAIIAN COFFEE BEANS
ISLAND
PRINCESS
A Gift from Maui
MACADAMIA
POPCORN
CRUNCH
NET WT. 26 OZS.
ISLAND PRINCESS
TROPICAL
Gems
THE ORIGINAL
ISLAND
PRINCESS
HAWAIIAN PEARLS®
CHOCOLATE
COVERED
HAWAIIAN COFFEE BEANS
MILK CHOCOLATE
A Gift from Maui
NET WT. 6 OZS.
ISLAND
PRINCESS
CHOCOLATE
COVERED
MACADAMIA NUGGETS
ASSORTED CHOCOLATES
A Gift from Maui
NET WT. 6 OZS.
ASSORTED CHOCOLATES
Island Princess
MACADAMIA
BRITTLE ROYALE
Luscious
Macadamia Brittle
Smothered in Chocolate

ISLAND PRINCESS MACADAMIA NUTS

Island Princess creates a traditional handmade package with tied black string around the wrapper of a glass jar. An uncluttered brown label corresponds with the nut-colored tones.

ISLAND PRINCESS — Chocolate Macadamias, Brittle

Textural boxes of silver and bronze back a modern graphic of palm branches and a sunset. Tropical Gems is part of the front window with pictures of sliced fruit.

Macadamia Brittle Royale displays a close-up photograph of chocolates on a similar-colored box.

BEARITOS

Organized freshness is emphasized with transparent corn chip bags which provide full view of the contents. The remaining products are bestowed with tones of yellow, brown and red which are typical of the Southwest.

"Little Bear," donning a sombrero, appears on all of the company's items. An earth-toned desert setting matches the colors of the product.

BEARITO'S POPCORN

Bearito's underscores their "organic," no additive or preservative quality with a product in clear plastic. The brown labels offer good contrast with the off-white popcorn.

BEARITOS
TORTILLA CHIPS
BLUE CORN
TACO SHELLS
TOSTADA SHELLS
REFRIED BEANS
Little Bear
10% LESS OIL THAN CORN CHIPS
ALL NATURAL NO PRESERVATIVES
STAYS FRESH TASTES GREAT
100% Organic Corn No Salt
BEARITOS
Little Bear
Organic Foods
POPCORN
TRADITIONAL
Ingredients: ORGANIC POPCORN Unhydrogenated Corn Oil, Butter, Natural Flavors, and Sea Salt.
Net. Wt. 4 oz. 113 g.
Little Bear Organic Foods
15200 Sunset Blvd., Suite 202
Pacific Palisades, CA 90272

MIGUEL'S
Tortilla Chips
Lightly Salted
Ingredients:
Corn, Water, Corn Oil, Salt
NET WEIGHT 8 OUNCES
MIGUEL'S
Tortilla Chips
NO SALT ADDED
MIGUEL'S
Mild Salsa Cruda

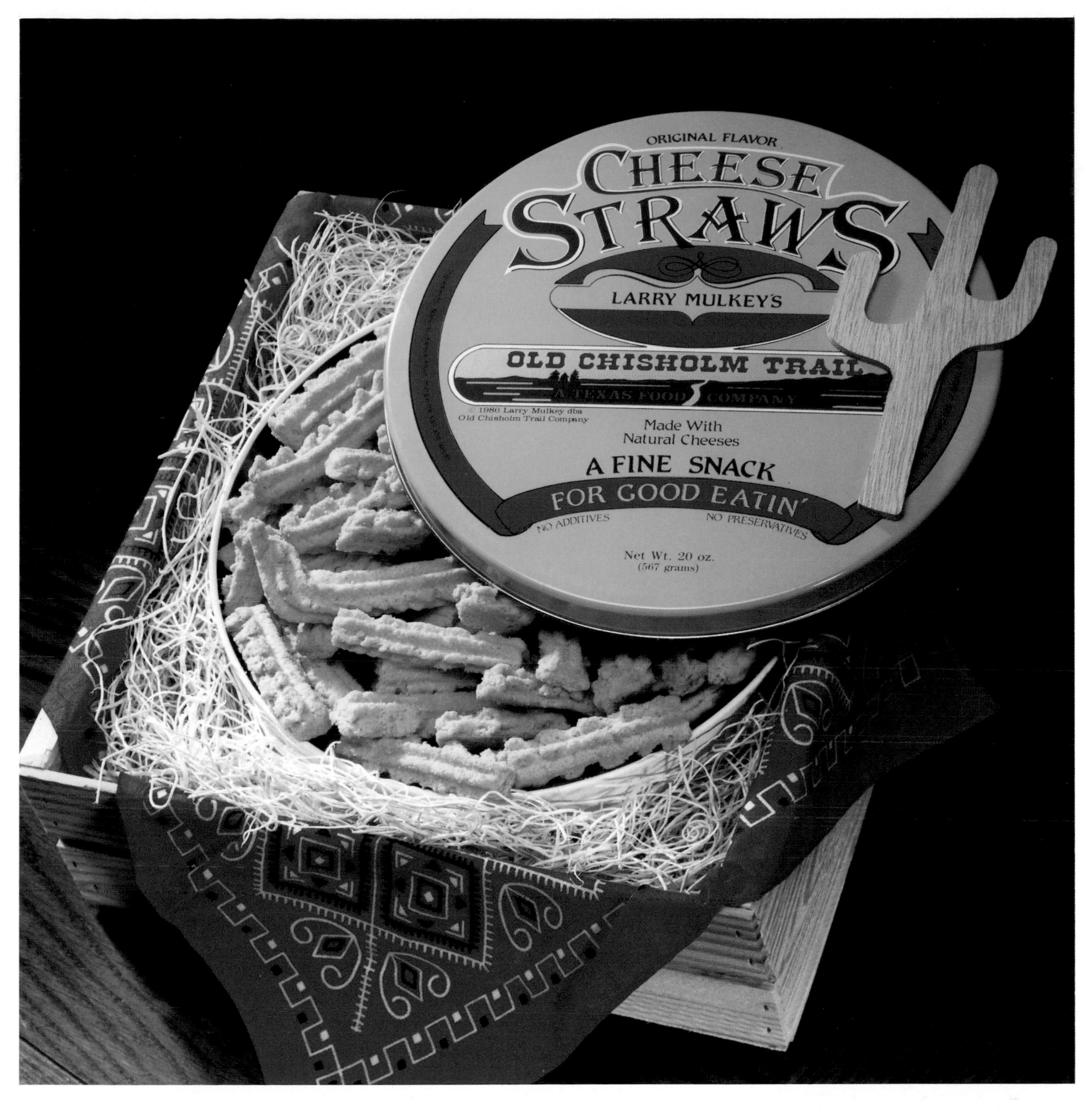

LARRY MULKEY'S CHEESE STRAWS

Multiple layers of packaging allude to a methodically homemade process. The metal canister is packed in straw (a homage to the product itself), draped in a western bandana and placed in a convenient wooden box. A narrow, horizontal illustration depicts the "Old Chisolm Trail" for additional ambiance.

MIGUEL'S TORTILLA CHIPS — And Salsa

Bright tones of yellow and blue evoke the Mexican origins of tortillas. The country's culture is further evoked with vegetables and seasonings pictured beneath an adobe home flanked by cactuses.

EDEN CHIPS

Eden Foods' chips are easily seen through a transparent bag. Asian authenticity is inferred with a red-and-black symbol over a traditional rendering of a bird in flight.

EL PASO — Coyote Nuts/Chili Fixin's

A rope illustration appearing at the edges of the box and a background desert scene evoke scenes of the Old West. The large image of one cowboy addresses the portion as a single serving.

Coyote Nuts take a humorous tack with a coyote caricature and catch phrase, "They're Howlin' Hot," at bottom.

MINA FRUIT SWEETS

Attractive geometrics enliven a wooden crate of Mina's Fruit Sweets, as various width parallel rows of fruit stand behind a centered lemon slice.

Fruit sweets®
HIGHEST FRUIT QUALITY
by Mina

LUKE'S ALMOND ACRES — California Nuts

Wooden crates of California nuts offer an enticing view of contents. The Company's emblem—three trees—is highlighted on a gold sticker.

ROYAL MEDJOOL DATES

Royal Medjool garners a "royal" reputation, visually, with gold-embossed dates hanging from a tree. Corner window allows for view of individually wrapped dates.

SMOKY VALLEY SOY NUTS

Smoky Valley's purple-on-white label stands out on a freshly sealed bag. A depiction of Kansas grain fields elicits a farm-grown impression.

WELL DONE POTATO CHIPS

The yellow-and-blue package suggests a Hockney-influenced painting with its freestyle approach. Brick-toned pyramid shapes and the yellow "smoking" effect conveys chips that are "Well Done," indeed!

CRAIGSTON CAMEMBERT

The logo of grazing cows is situated on the label above an elegant foil wrapper. A romanticized informational tag is attached with red ribbon.

Chapter Four

BEVERAGES
COFFEES
TEAS

Coffee has been America's most popular beverage since the 18th century. Nearly one-third of all world imports now come to the United States. For the discriminating customer, the choice is no longer limited to instants and canned varieties. Specialty whole bean coffee shops have from 30 to 120 varieties, offering a custom blend of beans.

The market for gourmet coffee continues to climb significantly despite price increases and concerns about caffeine. Each year since 1985, sales of whole-bean gourmet coffees have grown about 30 percent, with flavored and decaffeinated segments accounting for half of that growth.

Seventy percent of all coffee is still enjoyed at home, according to the International Coffee Organization. The remainder is consumed in offices, places of work and at eating establishments, where more people are turning to specialty blends.

The Washington, D.C.-based Specialty Coffee Association of America estimates that there are about 10,000 domestic specialty coffee stores — a number that increases every year. This is despite the influx of new coffee products — flavored instants, microwavable coffee, water-processed decaffeinated instants — and the in-roads into the gourmet coffee market by corporate giants like General Foods.

Personalized service is the key to the specialty coffee business. Better taste and quality are its best selling points. Once a customer tries coffee that's been brewed from fresh roasted and fresh ground beans, he's not likely to go back to stale beans and canned ground coffee.

Some stores build clientele by keeping a file on each customer's special blend, so that their "usual" can be quickly prepared and blended. Regardless of cost, it's a specialized form of attention.

Competition, accordingly, is stiff. Daniel C. Cox, co-chairman of the SCAA, has seen the growth. "The once exclusive coffee store has watched competition spring up everywhere as more and more retailers enter the market."

Specialty coffee roasters use three methods of distribution: retailing directly to customers, selling the product to specialty food stores, or distributing via direct mail. Supermarkets, which once stayed away from specialty coffee business, are now purveying private brands and installing in-store grinders.

Today's coffee and tea retailers agree that the big growth area in coffee is the specially-flavored varieties which are popular among younger consumers. The current best-sellers include hazelnut, amaretto, vanilla nut creme and chocolate raspberry.

For coffee lovers concerned with health issues, moderation is the rule. Research findings released in 1987 attested that steady coffee consumption is not as harmful as previously thought. Other recent studies say coffee consumption can improve performance and acuity at work, particularly in tasks that call for clarity of thought.

The surest way to sell coffee is to let people taste it. On-premise roasting also conveys a powerful image of freshness to consumers.

In San Diego, a full-service outlet uses a sales approach that combines specialty-store services with the convenience of a supermarket location. As the store roasts beans on the premises, the irresistable aroma draws customers throughout the supermarket.

Marketing experts caution retailers against stocking a lot of flavored coffees because oil flavored essences that coat the beans evaporate rapidly. Less costly, flavored coffees use an alcohol-based coating that disappears even more quickly. The retailer must be confident that the flavored blends purchased from the supplier are fresh enough to have a shelf life of three to four weeks.

Another concept in coffee marketing is the use of new packaging systems. More outlets are selling gourmet coffee in bulk, using new one-way valve bags. With one-way valve packaging, suppliers can distribute "fresh roast" coffee. Others are down-sizing their bags to improve freshness.

"Packaging is the product today," explained Bob Shedlock, vice president of marketing at New York's Hena Inc., in an interview with *Gourmet Today*. "People aren't selling coffee anymore; they're selling a coffee concept with very elaborate packaging. If you want to get a decent share of the market . . . , you have to come out with interesting new concepts of packaging."

A gourmet coffee center requires only eight square feet of space, while a new store or coffee department usually sells about 50 pounds per week. For a start-up operation a minimum of seven types of coffee is required. Experts recommend such standards as Columbia Supremo, Mocha Java, a special house blend, decaffeinated, French Roast Columbian and flavored Chocolate Almondine.

New outlets also need display devices. These can either be wooden barrels that hold 25-pound sacks of coffeee (for a vintage decor) or transparent lucite display bins (for a contemporary decor). Other accessories include a grinder, a scale of some sort, coffee scoops and bags for the ground coffee, preferably imprinted with the store logo.

For some specialty shop operators, a bulk tea sideline can increase the profitability of the store. Bulk tea is a natural adjunct to whole-bean coffee. Explained a New York-based tea and coffee company owner: "Basically, coffee is a lot more competitive than tea. There are less people selling bulk teas. With coffee, everybody on the street is selling it now — all the grocery stores and every corner deli. Selling tea gives us our own niche in the market."

Retailers typically buy bulk tea in 100-pound chests and display the product in the same containers. Customers usually buy tea in quarter-pound or half-pound orders, since a whole pound of tea is enough to make 200 cups, whereas coffee is typically purchased in one or two pounds. The most popular varieties of tea are English Breakfast, Earl Grey and Indian Darjeeling. Most people don't know much about tea, according to one shop owner, so they like to stick with something safe.

Camellia sinesis, the tea plant, is grown in plantations from seed and, more recently, from cuttings of selected clones. It is harvested by plucking the newest and most tender leaves from the tips of branches. The major tea-producing regions in the world are India, Sri Lanka, Indonesia, Africa, South America and China.

Although there are probably 1,500 varieties, all leaf teas from the tea plant can be classified into three catgories based on the way the leaves are processed: black, oolong and green.

Black tea, the most popular type, is fully fermented. Premium black tea varieties include the Indian-grown Darjeeling, Assam from northeast India and Pakistan, Ceylon and Sri Lanka, and Oolong and Keemum from China and Taiwan. Tea specialists identify premium black tea with the label "flowery pekoe" or "broken orange pekoe."

No two tea harvests are of identical quality, experts say. The harvests vary greatly because tea is extremely sensitive to variables in soil, climate and elevation. As a result, the burden of maintaining quality rests on the middlemen in the tea trade — the processors, brokers, buyers, tasters, blenders and packagers.

But unlike other specialty foods, tea must be carefully evaluated across all crop sources, and each variety must be continually blended and reblended to maintain the quality and flavor customers expect. Other factors affecting quality are where the tea was grown and when it was picked. As for blended, scented and flavored teas, the quality is determined by the skill of the blender, taster and vendor.

Like coffee, tea has suffered somewhat from the "bad press" associated with stimulating hot beverages. A medium-strength cup of black tea contains about one-third or one-half the caffeine of a comparable cup of coffee. Consequently, many vendors now offer a variety of decaffeinated blends. Concern over caffeine has also led to an increased interest in herbal teas, most of which have no caffeine. No longer restricted to health food stores, herbal teas sell in supermarkets and chain restaurants around the country.

Other contenders in the decaffeinated market are the green teas. Unlike black and oolong teas, green tea leaves are unfermented and contain no caffeine.

"New Age" soft drinks have also found a niche in the beverage market. More and more adults prefer these flavored seltzers and natural soft drinks over colas and sodas. The New York-based resaerch firm, Packaged Facts, reported that 1987 sales of New Age soft drinks topped $1.9 billion. Bottled waters and seltzers spiked with fruit essences are sold by many manufacturers, while a few companies offer fruit juice sparklers, or carbonated beverages of 0 percent to 90 percent fruit juices.

VERMILLION RED BLOODY MARY MIX

The Vermillion family crest is emblazoned twice on this upscale Bloody Mary mix. The use of red, black and gold colors heightens the bottle's dramatic impact.

TRUMP'S BREAKFAST ASSORTMENT

A large octagonal box provides immediate originality for Trump's Breakfast Assortment. Teas, scone and jellies are displayed on the box top, as are a classic blue teapot, tasteful cloth napkins and a beautiful wicker serving tray.

Tea from Trumps
Trumps
an invitation to Tea
Trumps own sweet pickled orange slice
Trumps
Trumps own Scone Mix
ounces of tea
Trumps own sherry wine jelly
Trumps own raspberry jam

TRADITIONAL CLASSIC TEAS

Illustrations of quaint teapots and table settings connote the gourmet atmosphere which characterizes Traditional Classics Herbal Teas.

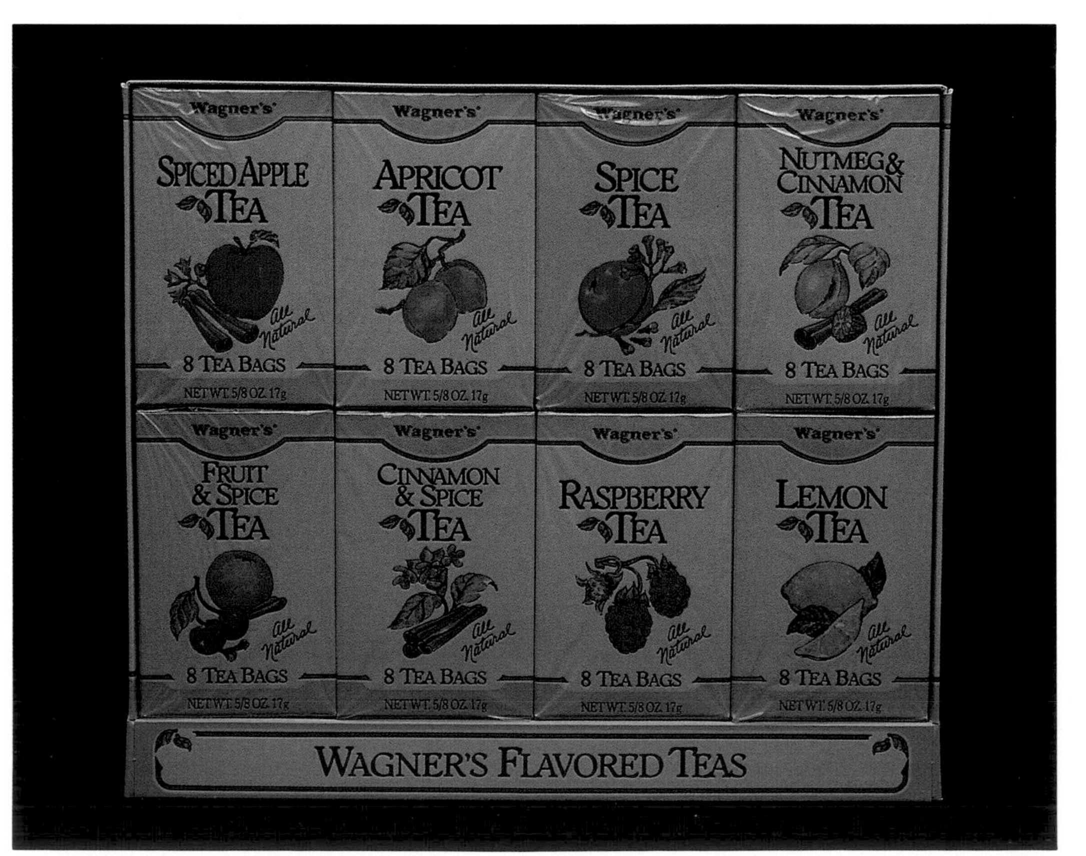

WAGNER'S TEAS

Wagner's Teas are available in an eight-pack which provides a visually wide selection of the product. The earth-toned boxes are differentiated by eye-appealing drawings of fruits and spices.

EASTERN SHORE TEAS

Decorative illustrations of Eastern Shore's various tea ingredients are color-coordinated with quaint ribbons situated over the fold of the bag.

EDEN SOY BEVERAGE

This natural soy beverage is dignified by an absolutely tranquil impressionistic scene. The peacefulness of the label's subtle hues mirrors the healthful intentions of the beverage.

COACH FARMS GOAT'S MILK

Packed in a convenient four-pack carton, Coach Farms Goat's Milk comes in white containers with contrasting red caps, alluding to wholesome dairy freshness.

JABLUM BLUE MOUNTAIN COFFEE

Jablum's Blue Mountain Coffees are logically saddled with blue-dominated labels. A soothing visual effect is attained through crisp, easy-to-read type and attractive illustrations of just-picked beans.

100% Blue Mountain Coffee
100%
Vacuum Packed
For Freshness
JABLUM
Jamaica Blue Mountain Coffee
PRODUCT of JAMAICA
Preground for your
convenience.
Vacuum Packed
For Freshness
Mountain Coffee

ALPENGLOW CIDERS

These upscale ciders replicate fine wines with tall bottles, fancy labels and sheer sophistication. The cider is identified graphically with an apple which is substituted for the letter "o" in Alpenglow.

SOUTHERN TOUCH MUSCADINE JUICE

Brightly colored labels spotlight appealingly sketched leaves and fruit. An easy-to-grasp bottleneck with a textured surface makes for easy handling.

BOYD'S SPICED CIDER

A black-and-white rendering of apple-pickers lends an old-fashioned touch to this cider. The neutral gray box is offset by a modern, yellow-red-and-blue stripe on the label.

DANESI COFFEE

A dark brown label coincides with the rich tones of Danesi's espressos.

DANESI COFFEE BEANS

The "D" logo on the face of the cup stands apart from the distinctive black or gold bag of coffee beans. A protruding valve guarantees the fresh quality of the merchandise.

CAFE TIERRA, COFFEE BEAN INTERNATIONAL, SUNDAY COFFEE

Glossy black bags are trimmed in gold to signify the decided sophistication of these gourmet coffees. Coffee Bean International and Sunday Coffee assure their sealed-freshness with pressure valves, while the Cafe Tierra package illustrates a vividly colorful mountain range.

GOTEBORG'S ASCOT CAFE

Gold type corresponds with gold wrapping on right, while silver corresponds with silver wrapping at left. A stately outer label couples a decorative place setting with a single flower.

ALLEGRO FINE COFFEES

Glossy paper bags indicate the premium nature of the gourmet coffee. White copy stands out clearly below a photograph of the dark beans.

BARRIE HOUSE BUZZ COFFEE

Barrie House produces this item with twice the caffeine, as exemplified by the word "Buzz" printed in a shock-like double typeface.

LES CAFE'S ORIENT EXPRESS COFFEE

A rich, yellow metal container portrays romantic settings on the front and side of the package. An affluent young couple—typical of the product's preferred clientele—is pictured on the famed Orient Express train.

DAMMAN SPECIALTY TEAS

The sophisticated use of a black and gold metal container is punctuated with traditional red, yellow and white Asian illustrations.

CASINO DE FRANCE COFFEES

The focal points of the packaging are centrally drawn scenes which coincide with a specific blend of Casino de France coffee. The red-and-green Espresso box presents a gondola in a Venice-like canal; the deep brown Moka box expresses a modernistic desert theme; and the 100% Columbian blend places an old-style sailing ship on an expensive-looking black box.

ISLAND PRINCESS COCONUT CREAM KONA BLEND COFFEE

Island Princess maintains consistency in their package design with a coconut cream-colored label. The informal typeface and subtle drawing express Hawaiian tranquility.

GHIRARDELLI'S — Ground Chocolates and Cocoas

Ghirardelli's brown containers depict an eagle touting "Premium Quality from San Francisco." The package of Ground Chocolates portrays a steaming cup on a red background.

FINLEY LTD. COFFEES

A regal woven pattern and a classic sailing ship connote the gourmet nature of Finley Ltd. Coffees.

GRACE RARE TEAS — Irish Breakfast

Dramatic calligraphy shares the package face with a classic sailing ship. The deep-red type stands out smartly on the upscale black-and-gold container.

CAFE LA SEMEUSE

A steaming cup of coffee is shown rising from a presumably-Swiss mountain top to convey hand-picked freshness. The glossy yellow bag is accentuated by a shocking pink sun.

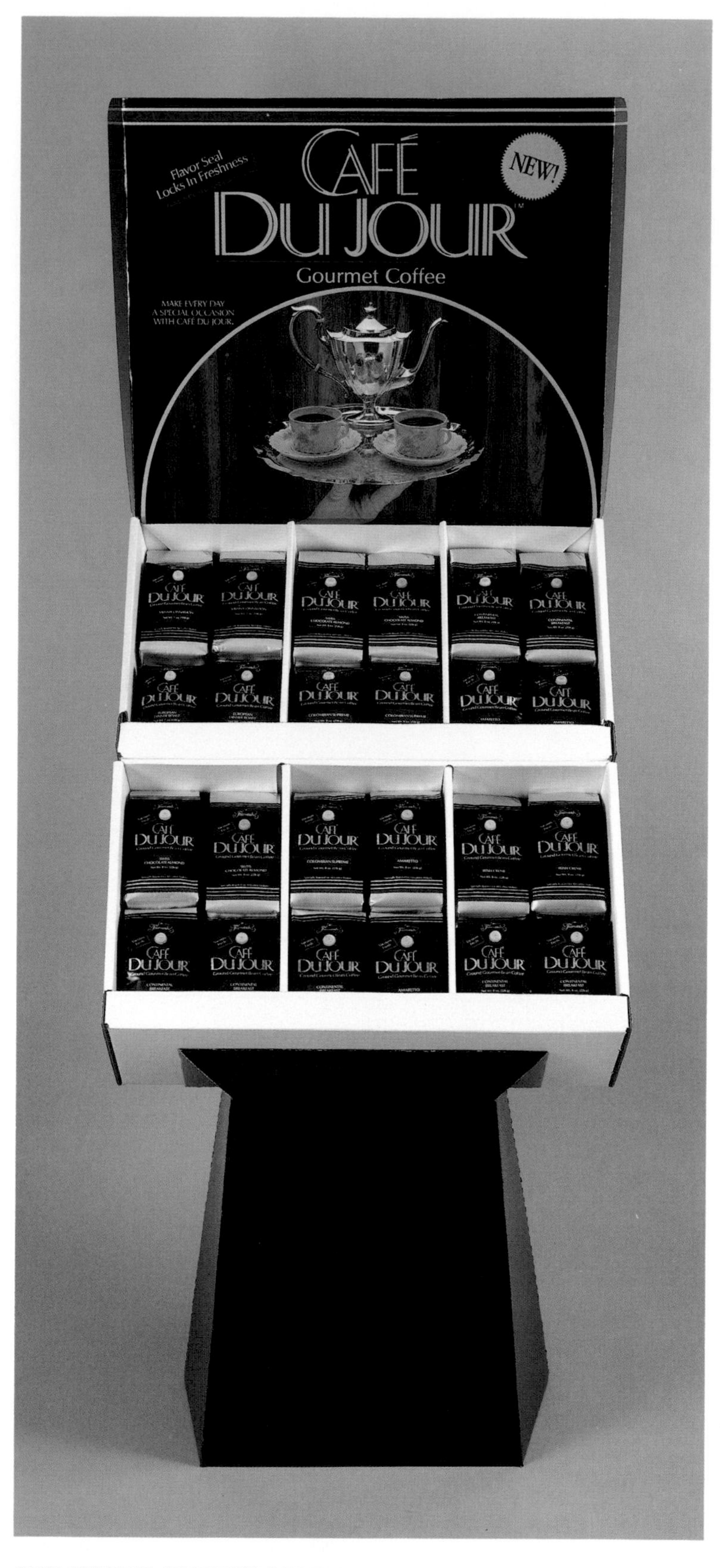

CAFE DU JOUR GOURMET COFFEES

A statement of understated elegance is made with a black, green-striped label. Well-defined copy makes clear the freshness and style of the merchandise.

CELESTIAL SEASONINGS TEAS

Intricately picturesque illustrations are worthy of further inspection, as is the gourmet product itself. The boxes only employ deep, warm-hearted hues for a friendly, soothing image.

FAIRWINDS GOURMET TEAS

Fairwinds' boxes duplicate the shape of a tea bag, complete with an attached "dunking" string. This gives the formal package immediate identification by shape alone.

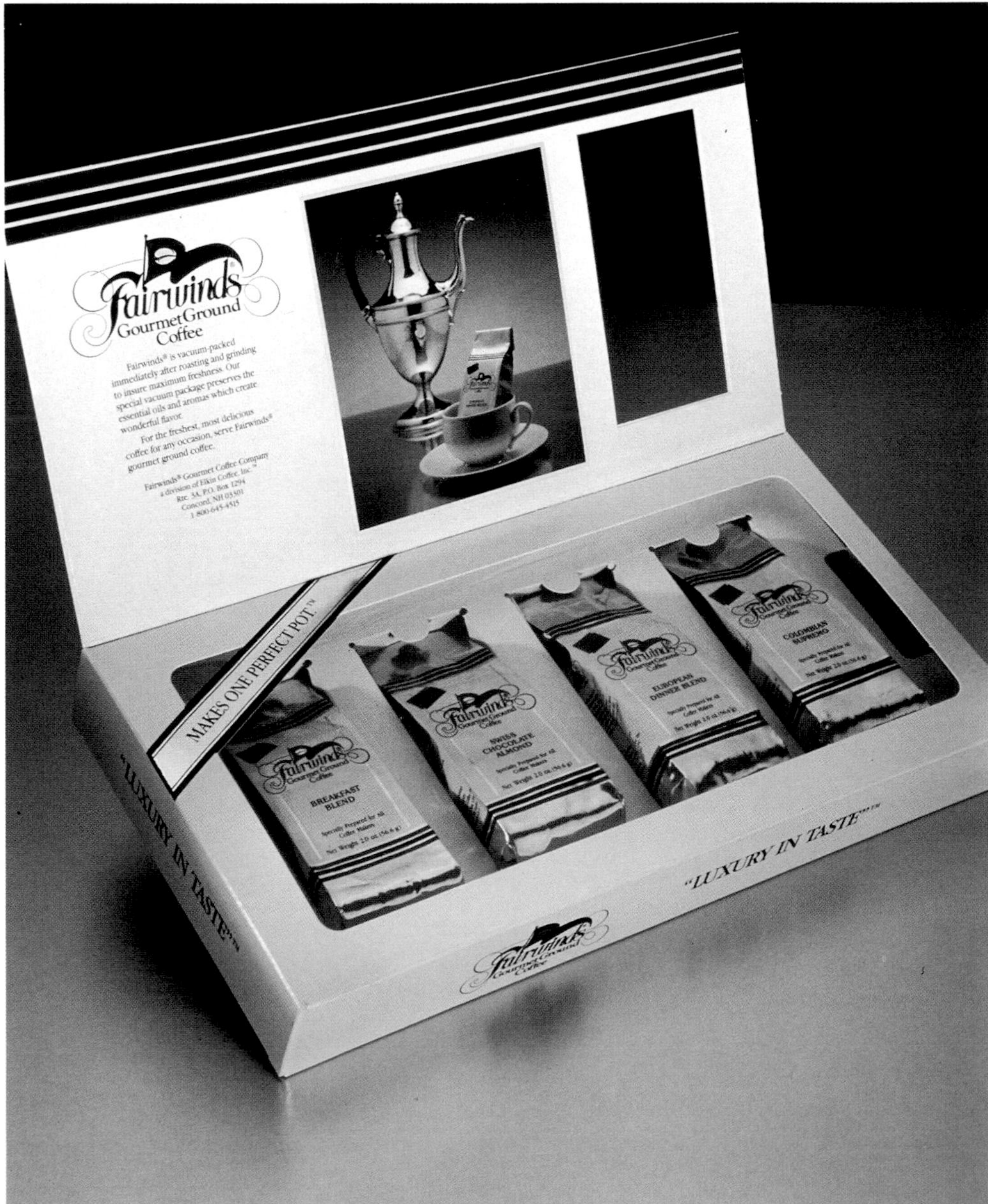

FAIRWINDS COFFEES

"Fairwinds" is embellished with ornate black lettering on a contrasting label. The classy gold bag is a vacuum-packed testimony to the coffee's premium quality.

Fairwinds
Gourmet Ground
Coffee
EUROPEAN
DINNER BLEND
Specially Prepared for All
Coffee Makers
Net Weight 2.0 oz. (56.6 g)

Chapter Five

PRESERVES JELLIES JAMS

To sweeten or not to sweeten. That's the question faced by makers of jams, jellies and preservers. Sugar-free spreads, exotic fruits and liqueur-laced preserves have changed the flavor of traditional recipes. More manufacturers now offer reduced-sugar fruit spreads, or "spreadable fruits" made with no refined sugar at all. Less sugar, fewer calories.

In the old days, jam and jelly were made by simmering fruit with sugar and a dash of lemon, then strained or reduced to a spreadable pulp. Times have changed. Today a multitude of fruits, berries and vegetables are now inside jars of jam and jelly.

The jams, jellies and preserves market is projected to hit $1.2 billion by 1990 according to New York-based research firm Packaged Facts. Retailers report lively sales of unusual flavors, such as tropical pawpaw from Australia. Fruit butters have also branched out, with extensions such as plum and pear butters.

Cooking fruit with lots of sugar was a good way to preserve spreads before refrigeration. Whole fruits cooked in thick sugar syrup are known as preserves. Fruit syrups or strained fruit purees, boiled with sugar, are jellies. Marmalades are made from chopped fruit, cooked with syrup into a thick spread. Fruit simmered with sugar into pulp is jam.

The Food and Drug Administration requires jams and jellies to have a minimum sugar content in order to be labeled as preserves. If a product does not meet such requirements, it must be labeled a fruit spread. Some manufacturers substitute fruit juice for sugar although this can result in a diluted flavor.

Although traditional homestyle fruit spreads are gaining popularity, many families prefer low-sugar fruit spreads for their health. The fruit-only spreads, the reasoning goes, make for healthier snacks. Diabetics also appreciate today's wider range of fruit-only products.

Peanut butter producers are turning out their own raft of rich sweets. The list of new peanut butter flavors reads like a menu at an ice cream parlor: Rocky Road, Dark Chocolate, Honey Cinnamon and Vermont Maple Syrup. These manufacturers are excelling in the traditional children's market.

For the adventurous, unusual and exotic fruits, such as star fruit and olallieberries, are available for spreads. One company even makes preserves from spices such as gingeroot and allspice, recreating the tangy flavor of Jamaican sweet spreads.

Liquor has also joined forces with fruit preserves, in flavors that include amaretto with apricot jam, scotch with orange marmalade and strawberries with champagne. A more potent spread is found in preserves spiked with champagne, rum, cognac, kahlua or aged scotch whiskey. "A simple croissant," explains one retailer, "can be turned into a luxury by spreading strawberry and champagne spread." Even so, raspberry and strawberry are still acknowledged as the heart of the jam and preserves business.

Retailers laud sampling as the best way to promote jams and preserves. "When we sample something, we move twice as fast as the other products," says a Westerly, Mass. store owner. "So we rarely bother with advertising."

Packaging is crucial to the jam and preserve market. Patterned glass jars using colorful labels and country-style motifs enhance the appeal of a product, especially against strong competition. Some specialty stores offer gift packs in handcrafted wooden containers — crates, boxes, four-jar wheelbarrows and more.

One important development in the jam and preserve market is the shedding of its dessert-only image. Some producers promote vegetable-based jams and preserves — such as three onion marmalade — as condiments for meats. Preserves now mix with oils to produce tasty marinades and glazes. Acidic jellies — such as cranberry or currant — make excellent glazes when mixed with wine or poultry stock. Apple jelly melted with honey makes a fine glaze for baked ham.

There are many ways retailers can whip up the customer's appetite for more jams, jellies and marmalades; no-sugar jam or preserves as a low-calorie topping for yogurt; melted preserves with maple syrup for a truly fruity pancake topping; pear preserves with lemon juice as a dressing for fresh fruit salads; or three-onion marmalade as a pasta sauce by heating with sauteed garlic sausage.

CLEARBROOK FARMS
Oregon Blackberry
CLEARBROOK FARMS
OLD FASHIONED PRESERVES
Oregon Boysenberry
CLEARBROOK FARMS
OLD FASHIONED
• PRESERVES •
California Peach
CLEARBROOK FARMS
Chocolate
RICH DESSERT SAUCE
Michigan Red Tart Cherry
Blackberry Fruit Spread
NO SUGAR ADDED
Northwest Strawberry
OLD FASHIONED PRESERVES

CLEARBROOK FARMS

Clearbrook Farms offers old-fashioned preserves in stylized mason jars. The labels depict wholesome, delicious-looking fruit, hinting at what is inside the jars which come packaged in a keepsake wooden crate.

VOLUNTEER JAM

Charlie Daniels' Volunteer Jam raised money at a benefit concert this country singer gave to aid farmers. The jar labels show an illustration of Daniels playing his fiddle, while the cap patriotically depicts three white stars on a blue background surrounded by a red-and-white striped border.

LOLLIPOP JELLIES, PRESERVES, BREAD

Red-and-white tablecloth prints dominate the company's design theme, bringing to mind family-style wholesomeness. A subtly drawn Lollipop Tree stands out crisply against the simple white packaging.

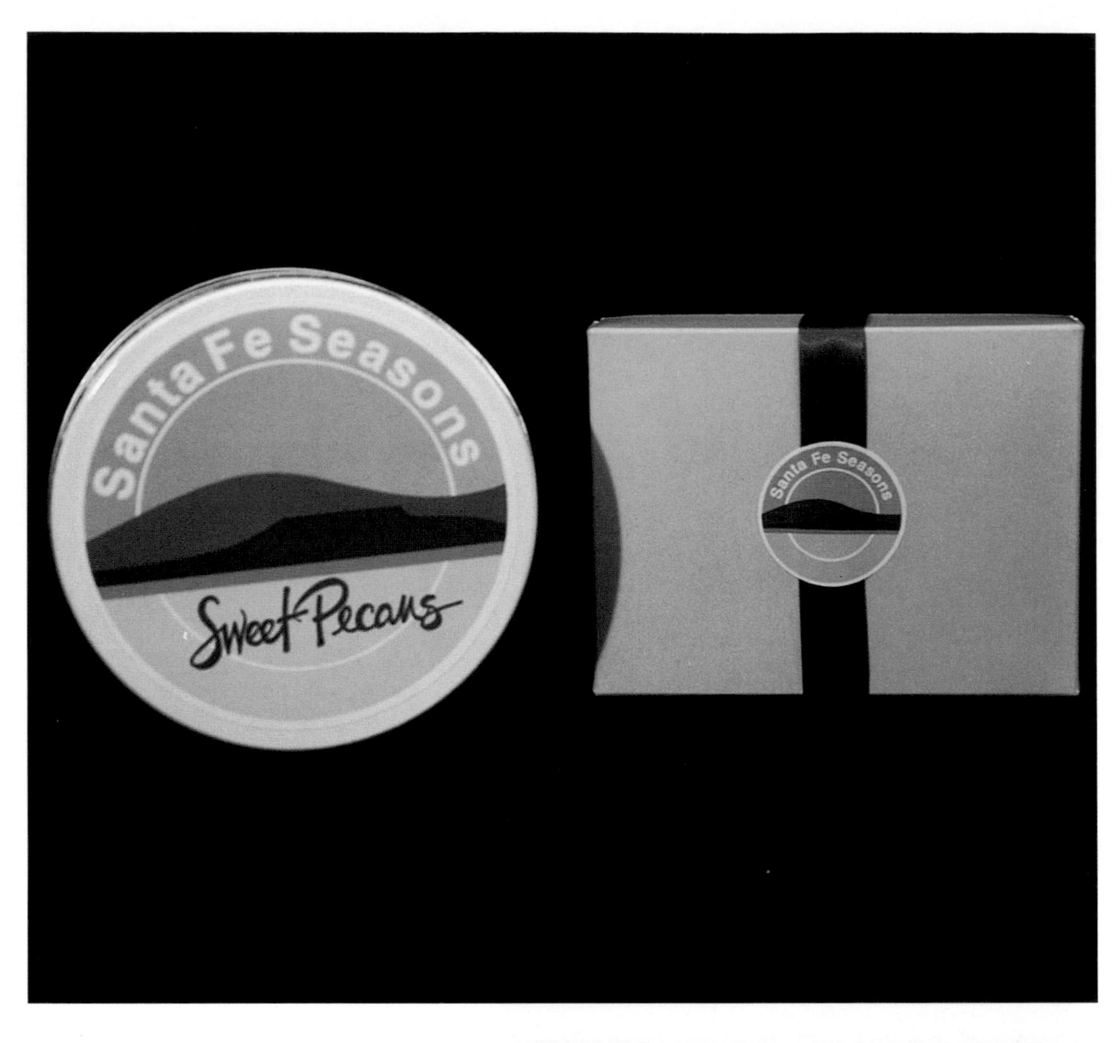

SANTA FE SEASONS

The logo and design for Santa Fe Seasons reflect many aspects of the Southwest. A zig-zag Indian motif and a circular logo which features a typical Santa Fe vista appear on a sand-colored background.

GRAYMARSH

The focus of the label for Graymarsh farm is the farm itself. The cattails add the perfect marshy feel.

SOUTHERN TOUCH JELLIES

The Southern Touch jelly labels share the same basic design as their syrups but are differentiated by the bottle shapes and the pictures of fruit on the labels.

SOUTHERN TOUCH SYRUPS

The labels for Southern Touch syrups are printed in earth tones and "syrup-like" colors.

OREGON HILL FARMS SYRUPS

The rich tones of these fancy fruit syrups glow like lamps as seen through geometrically sleek bottles. The black typeface contrasts opulently with the gold label and the long gold wrapping over the caps.

BERRY PRESERVES FROM IVYLAND

Vibrant caricatures of raspberries, blueberries and strawberries reflect the homemade attributes of the preserves. A gently dotted print on the box is similar in color to the product, and would look right at home as traditional kitchen wallpaper.

WAGNER GIFTBOX

The Victorian lithograph on the gift box lid makes the box worth keeping after the preserves are long gone and also functions well in a store display. The labels on the preserves themselves contain beautiful illustrations of the berries used to make the product within.

TRAPPIST PRESERVES

The monks of St. Joseph's Abbey continue a 1400-year tradition with their Trappist Preserves. Packaged in simple boxes, like the lives the monks lead, they are set in a typeface reminiscent of an illuminated manuscript.

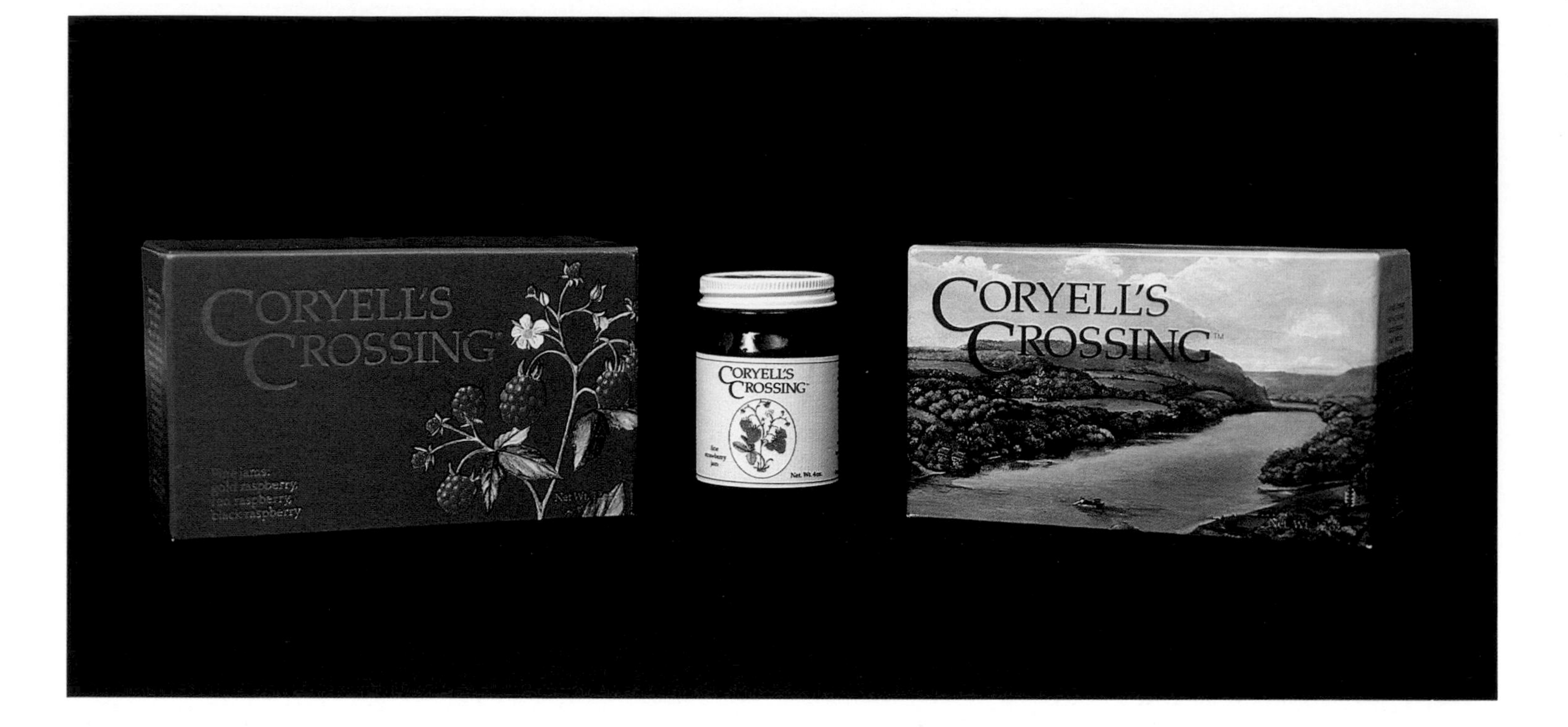

CORYELL'S CROSSING

A pastoral landscape graces the cover of one of Coryell's Crossing's boxes. Their award-winning raspberry preserves' box has an enlargement of the illustration which appears on the labels of the individual jars.

CORYELL'S CROSSING®
Net Wt. 10 oz.
280 grams
CORYELL'S CROSSING™
fine
raspberry
jam

d'arbo
Naturrein
KIWI STACHELBEER
MARASKA WEICHSEL
MULTI VITAMIN
GARTEN ERDBEER
WALD HIMBEER

d'arbo
Naturrein
Konfitüre
In Darbo Naturrein kommt nur Natur rein.
d'arbo
Naturrein
In Darbo Naturrein kommt nur Natur rein.

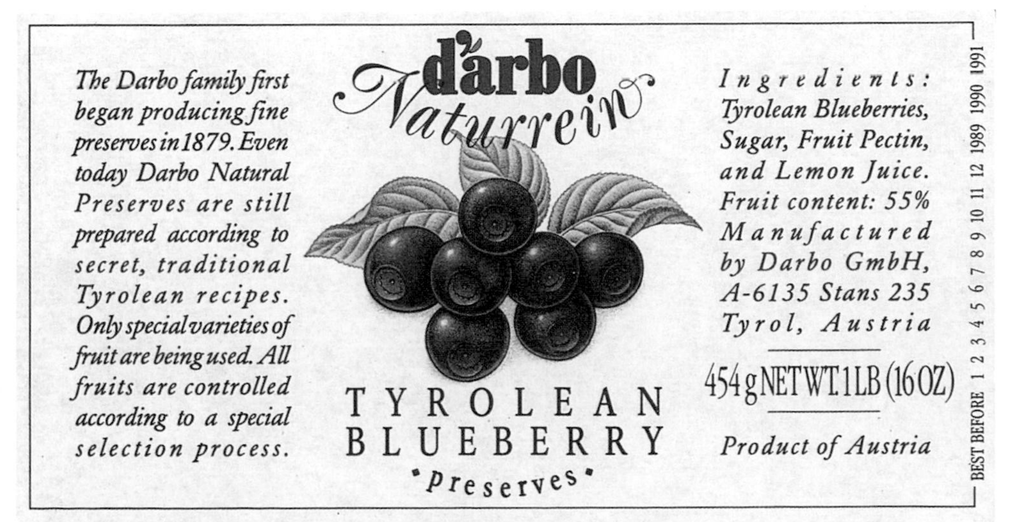

D'ARBO PRESERVES

Each of these preserves shows pride in its traditional Austrian origins with extensive text on the sides of the jars. Such historical references to the D'arbo family's century-old recipes evoke a trustworthy image. Tender illustrations of fruits, beehives and flower gardens add soothing hues to all the homestyle packaging.

BERRY BEST FRUIT

High quality illustrations appear as an inset oval on the labels of all the preserves from Berry Best Farms.

A native to the Mediterranean region, SWEET CLOVER now grows bountifully throughtout the hot, dry, high plains of the Western United States. Honey bees transform the nectar from the white and yellow flowers into a supremely unique, light honey.

Upon opening the jar, a cinnamon aroma announces SWEET CLOVER. Tasting reveals a spicy essence remarkably different from all other honeys.

These delightful characteristics have made SWEET CLOVER a honey most beekeepers speak of with joyous regard. We agree! The **Gourmet Honey Collection®** is pleased to offer this premium wildflower confection – SWEET CLOVER HONEY.

The **Gourmet Honey Collection®** invites you to experience fine honeys encompassing the cream of the crop from across the United States. The Collection showcases these national specialties so that you can savor each varietal flavor.

We handle only the finest in natural honey, insuring individual integrity and characteristics are retained. The **Gourmet Honey Collection®** process gently warms the honey, strains it through a fine screen and then bottles each jar carefully by hand.

Pure honey, like ours, may crystallize in time. Many honey connoisseurs select this creamy form because it is so easy to spread. Crystallized honey may be returned to a liquid state by placing the jar in hot (not boiling!) water.

For a real adventure into the wonder of nature's finest sweet, be sure to try all the different honey flavors packaged by the **Gourmet Honey Collection®**.

Distributed by: MOON SHINE TRADING COMPANY ®
PO Box 896, Winters CA 95694, ©1983

HIGH PLAINS SWEET CLOVER — Gourmet Honey Collection

This Gourmet Honey Collection has a sensual floral label which evokes the natural appeal of gourmet honey. Various intermingling pastels comprise the typeface of "Gourmet Honey Collection."

CINNABAR CHUTNEY — Marinade and Grilling Paste

Traditional Indian products are available in glass jars which provide a teasing view of the contents. An eye-catching beige label makes the red type stand out, and retains a reflective look.

Chapter Six

SALAD DRESSINGS
CONDIMENTS
SEASONINGS
VINEGARS
SAUCES

Ethnic, healthful and hot describe the top-selling products in the $3 billion condiments industry. As vendors introduce new products, consumers are expanding their preferences to spicy mixes, regional varieties and natural ingredients. Sales of condiments such as Mexican salsa, Creole mustard, Cajun spices and low-sodium salad dressings have climbed in the past two years.

Consumers are sampling innovative products. Cornichons, those tiny gherkin pickles, are natural garnishes for patés and sandwiches. Garlic has diversified with the birth of products like spiced marinated garlic. Exotic chutneys are showing up on retailers' shelves, too. The traditional mango chutney has given way to intriguing varieties. One store in Oregon offers Garden Tomato for grilled chicken, fish, steak and eggs; Plum curry for beef or lamb brochette, and Cranberry Occasion for veal, lamb and turkey.

Retailers no longer restrict their wares to store shelves. They now pack them in gift baskets which might include avocado oil, mustards, sauces and dips.

According to Edward Ogiba, managing director for the market research firm, Product Initiatives, "Condiments are one of the fastest growing categories in terms of introduction." Food specialist Rita R. Rousseau, reports that in 1987 some 1,200 new condiments were introduced in the market. Mexican salsa recorded the highest growth, increasing annual sales by 25 percent, while low- and reduced-calorie salad dressings and hot Cajun and Creole sauces also made jumps.

In the mustard category, jalapeño and Creole flavors are the latest trend. Mustards have found a variety of uses: as a basting sauce for barbecuing meats, an ingredient in homemade sauce or as a unique condiment for grilled foods.

One store grades its mustard from mild to hot. Customers with sensitive taste buds might start with Dill Mustard, then move up to Sweet 'n' Sour or even Stone Ground Mustard. Those who prefer condiments that bite back would enjoy Jalapeño Gold Mustard or Pub Mustard, the hottest of the line.

As with most fancy foods, consumer health concerns have prodded manufacturers to cut back on the sugar and salt content of their products. When customers inspect salad dressings, sauces, preserves and glazes, they look for labels that say "all natural" and "contains no preservatives." Jams, jellies and preserves made of fruit only, without added sugar, are encountering virtually no price resistance, retailers report. Consumers who are resigned to cutting back on sugar are willing to pay a higher price for "natural" foods.

Dressing a salad in the '80s requires more than just a vinaigrette. It now means tossing in special seasonings and toppings in addition to traditional items like oil, vinegar, lemon juice, salt and pepper, and a handful of herbs.

Producers now promote alternative uses for salad dressing: as baking sauces for fish, marinades for poultry and seasonings for dips and steamed vegetables. Specialty dressings thus come in many flavors, including apple-honey mustard, raspberry yogurt, toasted sesame, ginger-soy, lemon chardonnay and creamy peanut.

"In recent years," reports Lucy Saunders of *Fancy Food*, "food stores have discovered a new type of salad dressing: the gourmet dressing."

Gourmet salad dressings are set apart by designer labels and price. Salad dressings made from wine, fruit essences, nuts, vegetables and herbs compete with old standbys such as Italian, blue cheese and French. Merchandising has also grown more savvy, as some of the best-selling dressings are found not on shelves but nestled among the romaine lettuce, spinach and other fresh produce.

Many of the new salad dressings have dipped into international cuisines for ideas and ingredients. Spicy hot Thai peanut sauces, for example, have been adapted into a salad dressing, as have soy and ginger sauce from Japan, and sweet-and-sour sauce from China.

Who are the consumers of salad dressing? Couples with older children in the home are the most likely patrons. The larger the household, the more bottles of dressing used in a month. "Few consumers, however, want to have more than three of four bottles of salad dressing at home," according to the market research firm, Packaged Facts. Manufacturers are thus luring more purchasers by promoting the products as useful for more than just plain green salads.

In response to health-conscious consumers, producers are marketing no-oil and low-calorie dressings. Salad consumption has increased by as much as 21 percent, say market analysts, while more than 200 companies distribute salad dressings nationwide. But many customers eschew salad dressings, preferring a squeeze of lemon or a splash of balsamic vinegar. Balsamic vinegar, from sweet white Italian grapes, can be made into a light salad dressing simply by adding dijon mustard until it is a creamy liquid.

Vinegar is emerging as the seasoning of choice for all kinds of cooking. More varied and interesting than lemon or lime, and as effective as salt for heightening flavor, vinegar sauces are gaining popularity.

Substitutes for oil, high-fat and high-cholesterol thickeners have improved substantially over the years. Guar gum, xanthan gum, other vegetable gums and modified food starches allow manufacturers to create creamy, thick dressings with no oil, egg yolks or cream.

Nonetheless, oils (notably olive) with vinegar are still gaining steam in the market. Olive oil has some health benefits. The availability of extra virgin oil, which contains less than 1 percent oleic acid when pressed, has a far superior flavor due to better harvesting methods. With no more calories than any other oil, olive oils are enjoying brisk sales.

Like wine, olive oil comes in a multitude of flavors depending on the variety of the olive, the mineral content of the soil, the climate and the processing. Italy remains the largest exporter, ahead of Spain.

Olives also make one of the best relishes. For specialty retailers, the challenge is to find authentic varieties that preserve the traditional, natural methods of cultivation and curing. In contrast to mass-marketed olives, specialty olives display variations in color, texture and taste. Most premium specialty olives are allowed to ripen on the tree rather than undergo the artificial ripening process of commercially-produced olives.

Caviar, considered the "king of specialty foods," has slightly relaxed its lofty image. Along with premium caviars from the Caspian Sea and Volga River regions, current selections include variations in color, grain size, texture, flavor and, most importantly, price. Caviar is made from delicately-flavored roe of certain large fish, such as sturgeon, whitefish, salmon, lumpfish or paddlefish. "The best caviar," says *Fancy Food*, "has a sparkle and a glitter, regardless of the color. This beautiful shiny texture should be accompanied with the right amount of moisture — not too dry, but also never sticky. Grains should always be clearly separated, with no broken eggs." Fresh caviar is a premium product, whether imported or domestic. "Fresh" actually means packaged, since fresh caviar is always packed in tins or vacuum-sealed jars.

VILLA NICOLA LTD.

Elegant and handy cruets in dramatic gift boxes provoke an interesting combination of color and understated beauty. A red-white-and-green band refers to the Italian origin, as do the colors inside.

ROWENA'S CAPTAIN JAAP'S

Wicker baskets and a red tablecloth motif convey the homemade charm of Rowena's product line. Captain Jaap's nautical theme is found with logos of sailing ships on some of the products.

ROWENA'S
CAPTAIN JAAP'S
JAM AND JELLY FACTORY
CAPTAIN JAAP'S
Barbecue and Cooking Sauce
CAPTAIN JAAP'S
Lemon Garlic Olive Oil
ROWENA'S
Lemon Curd
Wonderful Almond Pound Cake
ROWENA'S
CAPTAIN JAAP'S
JAM AND JELLY FACTORY
RECIPES For Adventures in Cooking
CAPTAIN JAAP'S
ROWENA'S
ROWENA'S
CAPTAIN JAAP'S

SAN-J SOY SAUCE

Japanese characters near the base of the bottle allude to the product's point of origin. San-J's stylized rendering of suns acknowledges the "Land of the Rising Sun."

SAN-J TAMARI

Sharply contrasting colors of red, white and brown contribute to packaging which is both simple and dramatic. The Asian influence is seen with the graphic on the label.

SAN-J ALL-PURPOSE SZECHUAN SAUCE

The red label is well suited to a hot and spicy Szechuan Sauce, as is the fire-breathing Asian dragon.

HOUSE OF TSANG

Bright visuals make a bold statement on House of Tsang's black labels.

Mandarin Marinade

An Oriental treasure. Chinese style herbs and spices in a savory blend of dark soy sauce and wine. Makes meat more tender and succulent.

Mandarin Marinade

For a delicious marinade, use 1 table-spoon for every 8 oz. of meat. Marinate for 15-30 minutes, longer for poultry.

Write for recipes –
House of Tsang, Inc.
P.O. Box 294
Belmont, CA 94002

House of Tsang

House of Tsang™

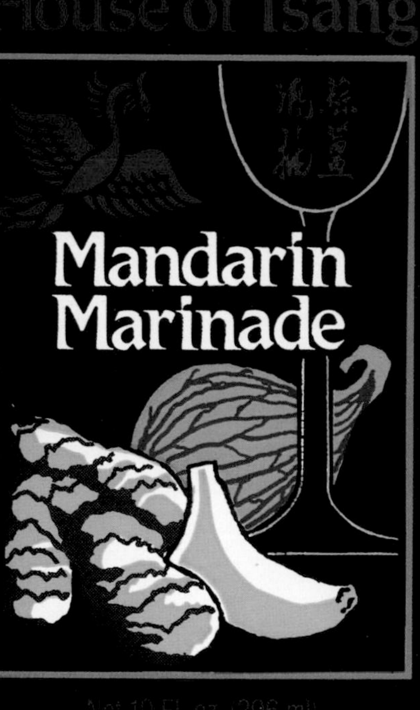

Net 10 Fl. oz. (296 ml)

How to Use

Use for marinating beef, poultry and sea-food. Brings a distinctive Mandarin taste to any meat. Also great for barbe-cuing. Brush frequently while cooking for added flavor.

INGREDIENTS: Soy Sauce (Water, Soybeans, Salt, Wheat, Caramel Color), Sugar, Water, Sherry Wine, Salt, Natural Flavors of Garlic, Ginger, Coriander, Onion and Black Pepper, Lactic Acid and less than 1/10th of 1% Sodium Benzoate as a Preservative.

NO MSG ADDED.

REFRIGERATE AFTER OPENING

0 75050 00605

Mfd. for
House of Tsang, Inc.,
San Francisco,
California, 94107, U.S.A.
Mfd. in U.S.A.

987

House of Tsang®

CLASSIC STIR FRY SAUCE
Transform any recipe instantly to a rich, distinctly Chinese Flavor. Stir-fry meat, poultry, or seafood with vegetables. Great as a table sauce for steaks, chops or leftovers for an Oriental Experience.

FOR STIR-FRYING: Heat wok or skillet over high heat with a small amount of WOK OIL®. Stir-fry your favorite combinations, add a dash of broth or wine if desired. Then add **CLASSIC STIR FRY SAUCE**. (1 T. per 1 cup of raw ingredients.) Toss to coat well and serve.

House of Tsang

INGREDIENTS: Soy Sauce, Sherry Wine, Sugar, Vinegar, Sesame Seed Oil, Cotton Seed Oil, Modified Food Starch, Salt, Dehydrated Garlic, Hydrolyzed Vegetable Protein, Dehydrated Onion, Xanthan Gum, Disodium Inosinate & Guanylate, Spice, and less than 1/10 of 1% Sodium Benzoate as a Preservative. (No MSG added.)

HOUSE OF TSANG, INC.
SAN FRANCISCO, CALIFORNIA 94107 U.S.A.
MFG. IN U.S.A.

Shake well before use.
Refrigerate after opening.

TRY OTHER HOUSE OF TSANG PRODUCTS:
OIL - SAUCES - SEASONINGS

WRITE FOR RECIPES
HOUSE OF TSANG, INC.
P.O. BOX 294
BELMONT, CA 94002

CHEF TELL'S PASTA POUROVERS

Chef Tell's illustrated likeness oversees his four jars of Pasta Pourovers. The corresponding ingredients to Sauce Suprema, Sauce Toscana, Alfredo Sauce and Salsa Verde al Pesto are pictured at bottom.

TRIOS

Anthony and Genevieve's all-natural sauces espouse their freshness with transparent jars, and indicate the type of sauce with a fitting illustration.

TAYLOR MAID — Blackened Spice Marinade

Specks of white on purple animate the "spice" in Blackened Spice Marinade, while a black maid at center dates the product to the Old South.

JOHNNY'S

Johnny's Chef-blended Seasonings employ a humorous quip to indicate which dressing is which.

PARCHED CORN HOLLER

The predominantly red label highlights this attractive premium sauce.

JOE'S SIERRA VISTA SAUCE

The illustrations of cacti reflect the spiciness of these all-purpose sauces.

FIREHOUSE

Robustly shaped bottles and fiery hues of red, black and yellow infer the spiciness of sauces that need a "firefighter."

CHEF ALBERTO LEONE

Not unlike the mirrors used during cooking demonstrations, overhead photos on each label attract discriminating gourmets to the merchandise.

TEXAS FRESH

The Texas Fresh family of products keeps a unity of design. Surrounded by a traditional Mexican border, each sauce is differentiated by a particular "South of the Border" color.

JUST DELICIOUS SOUPS, MIXES

What better way to imply the freshness of these soups than with pastel illustrations, white labels and plastic bags that give full view of the merchandise? Colorful ribbons which are color-coordinated with the drawings secure the labels to the bags.

JUDYTH'S MOUNTAIN — Garlic Butters, Sauce

The Judyth's Mountain labels foretell gourmet quality with dignified shades of beige, maroon and blue, all trimmed in glossy gold. Below the company emblem, a stately banner indicates the flavor of the sauce.

PUTNEY'S — Oil, Sauce, Chutney

A white background with delicate red and gray heart-shaped copy translates a healthy image from the packaging.

DEVOS LEMMONS — Belgian Classics

Devos Lemmon's Belgian Classics suggest a European origin with a landscape of an old castle and village.

NAPA VALLEY MUSTARD CO.

Simplicity of presentation and a sketch of the Napa Valley project the natural resources of that vicinity.

NAPA
VALLEY
MUSTARD
CO.

MAURICE ET CHARLES

The uncluttered Maurice et Charles label allows the contents to speak for themselves.

WALLA-WALLA

A simple sketch and clear copy make the bottled items sincere and easy on the eyes.

SONOMA TOMATOES — bits and chutney

Sonoma employs a vibrant red-and-green package scheme for their dried tomato bits and chutney, to remind consumers of the freshness of the product it contains.

SUCANAT

Consistent with their promotion of pure cane sugar, Sucanat (Sugar Cane Natural) boxes feature an illustration of a sugar cane plant, emphasizing their sole ingredient.

KOSLOWSKI FARMS

Gold and metallic tones on the labels provide a cool elegance for Koslowski Farms.

MONTANINI

Near perfection in product design is achieved through meticulously hand-packaged Italian foods. In stylized mason jars, olives, peppers, mushrooms and chick peas are so carefully arranged that eating the contents is almost an infringement on a work of art.

Tall glass jars, reminiscent of Roman columns, ask the question—how are the ingredients so neatly packed into such unique shapes?

GIAMBOLI

These uniquely shaped cruets of Montanini sauces are designed to be reusable and are equipped with a spout for easy pouring.

GOURMET DELIGHT

The gold materials which comprise the Gourmet Delight packaging reflect a pronounced prestige.

SINFUL FUDGE SAUCE

The colors of the distinctive label convey the richness of this high-end chocolate fudge sauce.

STEEL'S OLD FASHIONED FUDGE SAUCE

Steel's Old Fashioned Fudge Sauce comes in the traditional home-kitchen glass jar and is tied off with a piece of cloth to emphasize the time and care taken in preparing things the old-fashioned way.

QUAKER BONNET

An old-fashioned container equates well with the Quaker motif to portray the product.

Not just for Salads
COOK'S CLASSIC LTD.
AWARD WINNING PRODUCTS
COOK'S CLASSICS
COOK'S CAESAR
Dressing & Marinade
No Anchovies
TARCHON
Tarragon Herbal Dressing
Preservatives, No Sugar
LOW SODIUM — 35 mg per Tablespoon
COOK'S CLASSICS
Sheila Cook's
GARLIC LOVER'S
Dressing & Marinade
Low Sodium,
Preservatives, No Sugar and No MSG
375 ml
COOK'S CLASSICS
APPLE-HONEY-MUSTARD
Dressing & Marinade
Low Sodium,
No Sugar and No MSG
375 ml

COOK'S CLASSIC LTD.
AWARD WINNING PRODUCTS
COOK'S CLASSICS
GOLDEN ALMOND SAUCE
with a whisper of curry
No Preservatives, No Sugar and No MSG
SODIUM — 70 mg. per tablespoon
227 gm.

COOK'S CLASSICS HERBAL SAUCES

The ethereal-looking green label soothingly complements the herbal sauces.

OREGON HILLS — Fruit Sauce

Champagne-style bottles with foil wrappers are encased in a rural wooden crate. Topped with a holiday bow, the package is simple but elegant.

WAGNER SUNDAE SAMPLER

The five sauces in Wagner's Sundae Sampler are each labeled in a different color which denotes the flavor inside.

FRONTIER HERBS

Brilliantly colored seasoning packages parallel the fiery contents of herbs.

EL PASO CHILI CO. — salsa

The El Paso Chili Co. lassoes a western influence through cowboy-style rope around its labels. The cactus salsa has a green label to match a cactus illustration, while the salsa primavera suggests its contents with a pepper.

EL PASO CHILE CO. — Gift Package

The El Paso Chile Company's Gift Package is cleverly shaped and colored like an adobe home.

SNAKEBITE SALSA & EL PASO — Chile Con Queso

Snakebite's hot salsa — as the burning desert photograph implies — is "roped" for western effect, as is the chili con queso, whose light brown tone mirrors its mild taste.

WORLD OF SPICES

The gold label adds an air of prestige consistent with the exquisite taste of the spices.

TONY'S GARLIC — Bread Spread

The simple white label and lid of Tony's Original Garlic Bread Spread gives a perception of a family recipe. "Tony's" in script, comes across as a friendly signature.

EPICUREAN

What better way to present Saffron than with clear containers where color stands out amidst subdued background and copy?

SONOMA MUSTARDS

The passive hues of the Sonoma Coast packaging contrast with the tastes of the mustards.

FLAME

The barbecue sauce, pictured "wearing" a chef's toque, indicates its spicy flavor with a logo of a crowing rooster.

PAULA'S CALIFORNIA SEASONINGS
— And Oils

A central illustration is matched by color with each of the four herb seasonings—beef with a brown leather, fish an ocean blue, lamb with pasture green and poultry with a barnyard red.

Paula's Oils, in larger bottles, feature sedately colored labels and a simple flower graphic.

MRS. AULD'S — pickles, cherries, marmalade:

Mrs. Auld's stylized mason jars, adorned with pink or green ribbons, are made to look at home in any country kitchen.

DE NIGRIS

The straw-wrapped bottom of this unusually shaped cruet adds a well-placed dash of color.

HUNT CUP MUSTARD

A demure green label features an equestrian illustration, presenting Hunt Cup as a thoroughly high-end mustard.

COZETTE'S

Unobtrusive copy permits a good view of the product's richness.

FILOMENA

To portray a genuine old-world quality, Filomena's pasta sauce is packaged in large jars which resemble home canning. Moreover, gold foil labels carry Filomena's likeness for a personal touch.

GOLDEN WHISK —
Super Sauce

Save for a golden seal over, Golden Whisk is a clear bottle without a paper label, highlighting the appetizing red and orange hues of their sauces. A spoon and fork silhouette implies the culinary possibilities.

TROMBLY'S PEANUT SAUCE

Trombly's Peanut Sauce is appealingly affixed with a brown label which is similar to the color of the sauce.

ACROSS THE BORDER — chili, mustard, relish

Spicy southwestern seasonings are the heart of Across the Border's Jailhouse Chili, Relish and Jalapeño Mustard — signified by flaming red-and-yellow labels. The typeface is reminiscent of the Old West.

EDEN SHOYU SOY SAUCES

Smoothly shaped bottles of soy sauce suggest authenticity with a Japanese symbol framed by a circular background.

SELECT ORIGINS CULINARY OILS

Saute and Olive Oils by Select Origins project the distinctiveness of wine with tall, narrow bottles and wine-like labels. An illustration of a drop of oil is painted at the center, providing simple identification.

SABLE AND ROSENFELD

Sable and Rosenfeld's red cover immediately draws attention to the products. The same eye-catching pattern is repeated on the label in reverse color.

SOPHISTICATED NIBBLES — Dips

Sophisticated Nibbles' line of four dips presents a streamlined look with tall, thin boxes and light graphics. Illustrations of the ingredients located near the bottom subtly infer the contents.

COUNTRY SEASONS

The unusual shape of the Bayou Bar-B-Que Popping Corn container and its dramatic colors attract the eye from afar.

TRY ME

Striking visuals on each of Try Me's sauces are enough to entice the consumer.

GOURMET BUTTER & SPREADS
— Chocolate Crunch

A clean, unencumbered label gives the chocolate crunch an air of wholesomeness.

UPSIDE-DOWN CAJUN — pickles, chow-chow

Inside a keepsake wooden crate lay Cajun pickles and chow chow, with labels placed "upside-down" for a unique perspective.

JAKE'S — Barbecue Concentrate and Grilling Spice

In a wooden crate, a small hay bundle sits alongside Jake's Barbecue Concentrate, giving an outdoor feeling fitting to the product. Grilling Spice, in a similarly bold, modern jar, uses a green-and-yellow color scheme.

AMERICAN SPOON
— Cranberry, Plum Catsup

A gold and red logo is paralleled by a gold lid and the dark red contents — an attractive execution of simplicity.

AMERICAN SPOON — Dried Cranberries, Blueberries, Cherries

A dynamic view of hearty red cranberries, blueberries and cherries is readily appetizing through the transparent bag. A red-white-and-blue label expresses the product's American heritage, as does a colonial-era illustration on the boxed version.

CHALIF

Pastel tones on white recreate the smoothness of Chalif's Mayonnaise.

CHALIF
MUSTARD SEED
MAYONNAISE
100% NATURAL
CHALIF
FRESH GARLIC
MAYONNAISE
100% NATURAL
CHALIF
HONEY
MUSTARD
100% NATURAL
CHALIF
PERFECTLY PLAIN
MAYONNAISE
100% NATURAL
CHALIF
STRAWBERRIES
& CHAMPAGNE
MAYONNAISE
100% NATURAL
CHALIF
LEMON-DILL
MAYONNAISE
100% NATURAL
CHALIF
SESAME-CURRY
MAYONNAISE
100% NATURAL
CHALIF
CHAMPAGNE
MUSTARD
CHALIF
HONEY
MUSTARD
CHALIF
TARRAGON-ONION
MAYONNAISE
100% NATURAL
CHALIF
HOT & SWEET
MUSTARD
CHALIF
CHAMPAGNE
MUSTARD
100% NATURAL
CHALIF
COARSE & SASSY
MUSTARD
CHALIF
COARSE & SASSY
MUSTARD
CHALIF
STIR-FRY
SAUCE
100% NATURAL
CHALIF
HOT & SWEET
MUSTARD
100% NATURAL

ANNIE'S

Annie's label overflows with ingredients, implying that her sauces are similarly bountiful in taste.

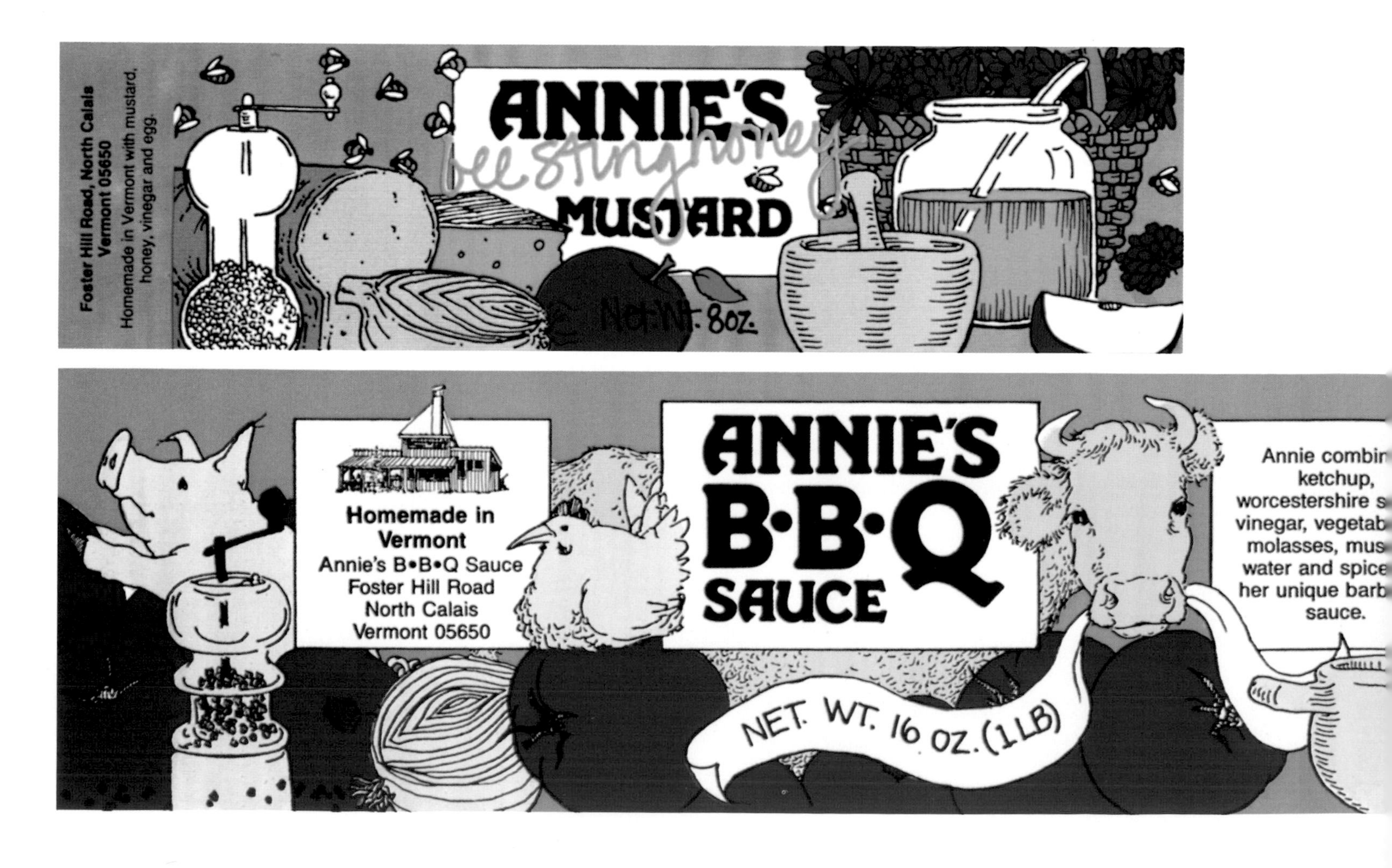

ANNIE'S
MUSTARD
Net.Wt. 8oz.
ANNIE'S
RASPBERRY
VINAIGRETTE
Homemade in Vermont
Net.Wt.8oz.(227G)
ANNIE'S
MUSTARD
ANNIE'S
SEAFOOD
SAUCE
Homemade in Vermont
ANNIE'S
B·B·Q
SAUCE
NET. WT. 16 oz. (1LB)
ANNIE'S
MUSTARD
Net.Wt. 8oz.

Chapter Seven

MISCELLANEOUS

Health foods are moving from the upscale markets to the supermarket scene. In California, frozen yogurt stands have proliferated in the past few years. In the cheese industry, low-sodium is the watchword. Trail mixes, once the choice of college students and fitness nuts, entered the grocery mainstream in the '70s and upscale market in the '80s.

Ready-to-serve meats have made their own mark in the snack food industry. In a 1988 NASFT survey, specialty food retailers reported the meat category as their number two bestseller. Specialty meats include fresh sausages, cooked hams, smoked poultry, pickled and canned meats. Lunch meats such as minced ham and smoked turkey breast are usually cured, then fully baked and cooked. Fresh sausage varieties have surged in consumer interest, as have smoked duck, pheasant, squab and quail.

For the highly perishable smoked fish, the greatest threat is spoilage. Retailers should keep smoked seafood no more than two weeks, below 36° F in a well-ventilated refrigerator case, according to the National Fisheries Institute in Washington, D.C.

The demand for paté has also risen. Today's paté ingredients run the gamut, from ground pork, veal, rabbit, chicken or duck liver, to salmon and vegetables. Pistachio, truffles, creams, prunes, raisins, and a number of liqueurs are new additions to patés. Madeira, cognac, port or other wines are essential to enhancing the flavor of the meats and spices.

Experts see a trend toward lighter patés. Vegetarian paté sales have grown as customers try to combine cravings for specialty foods with a healthy lifestyle. Smoked salmon, trout, cod, and seafood are also appearing as paté ingredients.

In terms of pounds consumed, pizza just might be America's number one food. Often just a phone call away, this meal-in-a-box provides a perfect dinner alternative. Pizza's recent popularity can be attributed to the successful marketing efforts of home-delivery giants such as Domino's Pizza.

Recent develops in frozen pizza technology provide an opportunity for specialty food shops and upscale supermarkets to add pizzas to their freezer shelves. And mass market players like Stouffer's and Pepperidge Farm, which introduced croissant and French bread pizzas, are reaping profits.

What snack food diet would be complete without cheese? Strong flavors like Brie and lush triple creams are showing brisk sales. The Saga Blue from Denmark—a white, mold ripened cheese with injected blue mold—is now being manufactured in the U.S.

Milk sources for cheese now include sheep milk, soy milk and goat milk. Health-conscious cheese lovers know that goat cheese, lower in fat and calories, is a good alternative to cow's milk cheese.

One of the NASFT's 1988 Special Recognition Awards went to Vermont Fromage Blanc, a fresh cow cheese made from skim milk and culture. Also taking hold are modified cheeses, some reduced in cholesterol and calories, others low in sodium.

Cheddar is the hottest-selling cheese, according to *Fancy Food*'s 1988 survey. In second place were the rich double and triple cream cheeses, such as St. Andre, followed by Swiss-style cheeses.

Italian cheeses are in demand. Parmigiana Reggiano, Fontina, provolone, and mozzarella are becoming familiar names. Predictions for the best-selling cheese of the 1990s? According to *Fancy Food*, "it will have the rich, buttery flavor of a St. Andre, with the fat content of a sapsago."

Creative packaging is one secret behind the snack food boom, as many retailers consider a label as important as the packaging. Many food producers achieve a simple, expensive look with quality paper and unusual colors of ink.

Gift packs are no longer limited to the usual crackers, candies, chocolate and cheese. In fashion are theme baskets for weddings, births, holidays and business gifts. One shop in Oak Park, Ill. has made a killing with one homestyle carry-out cuisine packaged in gift baskets. Its Coast to Coast theme basket contains regional American foods, including East Coast crab, boiled spices, dried sour cherries from Michigan, Minnesota wild rice and California olives.

The trick is to create something unlike the typical card-and-gift shop or florist's basket. The right color combination can quickly give baskets a personalized touch. In bags, boxes, crates and mugs, most of the foods in gift baskets are old favorites such as candy, jellies, jams, sausages, oils, vinegars and salad dressings. The more imaginative designers offer such creations as Country Kitchen baskets, Sante Fe Baskets, Barbecue Baskets and Pasta Perfect Baskets.

AMERICAN SPOON — Dried Cranberries, Blueberries, Cherries

A dynamic view of hearty red cranberries, blueberries and cherries is readily appetizing through the transparent bag. A red-white-and-blue label expresses the product's American heritage, as does a colonial-era illustration on the boxed version.

WOODLAND PANTRY
WILD MUSHROOM SOUP

A classy pastel sketch of sliced and whole mushrooms is the visual crux of the label.

WOODLAND PANTRY MUSHROOMS

Radiant rural scenes provide an exquisite aura. The box tops depict "Woodland" scenes, and the fronts of the packages render likenesses of the enclosed variety of mushroom.

ROTHSCHILD RED RASPBERRIES

Numerous abstract and geometric glass designs are esthetic gems in their own right, while opulent red raspberries in liqueur project rich hues behind the provocative company logo.

ROTHSCHILD — Oils, Vinegars, Preserves

These Rothschild products are uniformly packaged with incredible stateliness. Provocatively tall bottles of oils and vinegars are housed in gold-woven print boxes, while the preserves use a similar box in horizontal form. The company logo, in green and gold, displays the Robert Rothschild farm.

HOGUE FARMS
PEARS
HOGUE FARMS
PEACHES
NET WT. 14 OZ.
HOGUE FARMS
CHERRIES
HOGUE FARMS
APRICOTS
BEANS
Hot & Spicy
HOGUE FARMS
NEW!
SNAPPERS
ASPARAGUS
Pickled & Spiced
HOGUE FARMS

CHINOOK ALDER-SMOKED SALMON FILLET

A handcrafted wooden gift box houses Chinook's Alder-Smoked Salmon Fillets. An original cover design by an Alaskan artist invokes the tradition of the Alaskan Coastal Indian.

HOGUE FARMS — Fruits, Beans, Peas, Asparagus

Transparent jars expose invitingly attractive fruits and vegetables. Hogue Farms ascribes to premium quality with a freshness band sealed over its white lid.

ALASKAN SMOKED SOCKEYE SALMON

A picturesque dockside view, accented in a horizontal expanse, meets the pure Alaskan waters—such healthful attributes are fundamental to the Alaskan salmon's appeal.

ALDER SALMON FILLET

Meticulously drawn sockeye salmon, accompanied by dignified type on a white background, denotes a distinctly upscale item.

NOUVELLE SORBET

Regal hues of red and blue set the pace for the upscale, gold-trimmed sorbet container. The dessert's French heritage is acknowledged with delicate fleurs-de-lis on the label.

RICE DREAM

A sky blue container features images of "floating" ice cream cones to produce a light, dreamy ambiance.

GONE BANANAS

The catchy product name is evoked humorously with a caricature of a banana wearing sunglasses and carrying a suitcase. The bright yellow-and-blue color scheme is equally cartoonish.

SYBIL'S KITCHEN

Homey pastels of pink, green and peach differentiate the dinners from Sybil's Kitchen. Sybil's photograph appears next to a one-quarter view of the product on a plate.

LOOK'S ATLANTIC — Bouillabaisse Soup

The primitive drawing of a soup tureen and the ingredients used to prepare the soup ascribe to a homemade European quality.

FANTASTIC FOOD POTATOES

A dark dish of light-colored potatoes divides the package into two parts, the white portion of which contrasts with the vividly seasoned product.

JAKE'S CLAM CHOWDER

The unique geometric shape of the package allows for easy stacking, and the perforated handle for easy carrying. The typeface used for "Jake's" and the illustration of clams both harken to 1892, the year of the company's founding.

EDWARDS VIRGINIA SAUSAGE

The country origins of hickory smoked sausage are implied with a drawstring canvas bag.

EDWARDS VIRGINIA HAM

Traditional Virginia Ham is packaged in an old-style wooden crate which uses black typeface to simulate branded wood.

PERFECT ADDITION SOUP STOCKS

Soup stocks of chicken, beef and fish employ humorously drawn cartoon figures wearing glasses, earrings, tiaras and human expressions. The clear white containers attest to homemade freshness.

STEELTIN STOCK DESIGN CANS

Round custom tins are available in a wide variety of diameters, heights and esthetically rural illustrations. Wholesome scenes such as those by Currier and Ives, are genuinely delightful.

TROMBLY'S PEANUT BUTTER

A two-toned brown lid and a simple brown label reflect the product's "Fresh Roasted" process, while the yellow typeface is the color equivalent of the peanut butter.

JODY MARONI'S SAUSAGE KINGDOM

Brilliant blocks of color contrast aggressively with flat black labels.

BEAN CUISINE SOUPS

Each of five different soup blends are appealingly fresh in clear plastic bags. A tied blue ribbon matches the background of the logo of steaming soup.

THE GIFTED LINE — Floral Collection

Elegantly designed gift boxes, bags, wrap, tags and stickers all possess sensual floral patterns from an 1893 Victorian scrapbook. The black packaging features rich hues which convey an unadulterated sense of prestige.

SMOKY VALLEY GRAINS

Unique ceramic tops, which are interchangeable to match the season of the year, serve as a lid for a round, White Pine box. The ceramics and the boxes contains Smoky Valley Grains and are all handmade.

BOOMBALA BOXES

Gorgeously ornate bows are trimmed in gold or silver. Discerning buyers will be as impressed with the boxes as with any contents.

THE GIFTED LINE — Basket Boxes

For basket boxes, which meticulously illustrate heartwarming images, the medium is truly the message. Victorian and Edwardian-influenced designs imbue the packaging with unparalleled dignity and class.

THE GIFTED LINE — Classic and Mini-Bags

Haughty illustrations of personified animals, floral prints and touching holiday scenes are the thrust of the upscale packaging. A high-gloss finish contributes a delicate touch to an already graceful item.

index

Queen Bee Gardens 53
1863 Lane 11½
Lovell, WY 82431
(307) 548-2543
(800) 225-7553
Honey Essence Truffles

Rice Dreams 238

Richard H. Donnelly 46

Robinson Ranch Specialty Foods 189
P.O. Box 1018
Walla Walla, WA 99362
(509) 525-8807

Rothschild Berry Farm 232, 233
Urbana, OH 43078
(800) 356-8933
(513) 653-7397
Robert Rothschild

Rowena's & Captain Jaap's 101, 169
758 West 22nd Street
Norfolk, VA 23517
(804) 627-8699
Pound Cake, Preserves, Sauce

Royal Medjool Date Gardens 119
P.O. Box 930
Bard, CA 92222-0330
(619) 572-0524

Rubschlager Baking Corporation 91
3220 West Grand Avenue
Chicago, IL 60651
(312) 826-1245

Sable and Rosenfeld Foods Limited 217
20 Richmond Street East
Suite 610
Toronto, Canada M5C 2R9
(416) 366-0386

Sabra 54

Sahagian Associates, Inc. 58
124 Madison Street
Oak Park, IL 60302-4206
(312) 848-5552
Boundary Waters Yard of Bubblegum

San Anselmo's Cookies 70
P.O. Box 2822
San Anselmo, CA 94960
(415) 258-0743

San-J International, Inc. 170, 171
2880 Sprouse Drive
Richmond, VA 23231

Santa Fe Seasons 154
624 Agua Fria #2
Santa Fe, NM 87501
(505) 988-1515

The Seckinger Lee Co. 78, 79
P.O. Box 14263
Atlanta, GA 30324-1263
(404) 231-4039

Select Origins, Inc. 88, 216
Building 139, Suffolk County Airport
Westhampton Beach, NY 11978
(516) 288-1381
(800) 822-2092
Rice, Oils, Dip Mix

Shaffer, Clarke & Co., Inc. 98
Three Parklands Drive
Darien, CT 06820-3639
Carr's Muesli Cookies

Sierra Vista Corp. 178
P.O. Box 1046
Orinda, CA 94563
(415) 548-0675
Joe's Sierra Vista Sauce

Sinful Chocolate Sauce 196

Smoky Valley Grains 119, 249
RR #1 Box 19
Marquette, KS 67464
(913) 546-2752
Soy Nuts

Sonoma Coast Mustard Co. 206
P.O. Box 64
The Sea Ranch, CA 95497
(707) 785-2166
(415) 383-6847

Sonoma Tomatoes 189

Sorrenti Family Farms 65
14033 Steinegul Road
Escalon, CA 95320
(209) 838-1127
Sorrenti Rice

Southern Touch Foods 133, 156
2212 B Street
Meridian, MS 39301
(601) 693-0602
(800) 233-1736
Southern Touch Juice, Jam, Syrup

Splendid Chocolates 54, 55
4810 Jean Talon W
Montreal, Canada H4P 2N5
(514) 737-1105

Steel's Fudge Sauce 196
425 Hector Street
Conshohocken, PA 19428
(215) 828-9430
(800) 6-STEELS

Steeltin Can Corporation 245
1101 Todds Lane
Baltimore, MD 21237
(301) 686-6363
Stock Design Cans

St. Joseph's Abbey 159
Spencer, MA 01562
Trappist Preserves

Susie Q's Pinquito Beans 74
Route #2, Box 1018
Santa Maria, CA 93455
(805) 937-2402

S. Wallace Edwards & Sons, Inc. 243
P.O. Box 25
Surry, VA 23883
(804) 294-3121
(800) 222-4267
Edwards Ham, Sausage

Sweet Adelaide Enterprises, Inc. 125, 208
3457-A S. La Cienega Blvd.
Los Angeles, CA 90016
(213) 559-6196
Tea From Trump's, Paula's Oils

Sybil's Kitchen 240

Taliahaus 30

Taylor Oil Company, Inc. 176
810 Poindexter Street
P.O. Box 8841
Jackson, MS 39204-0841
(601) 353-3455
Taylor Maid Blackened Spice Marinade

Texas Fresh Sauces 181

Trade Marcs Group, Inc. 143
55 Nassau Ave.
Brooklyn, NY 11222
(718) 387-9696
Cafe La Semeuse

Traditional Classics 126

Trombly's PBF Inc. 45, 214, 245
64 Cummings Park
Woburn, MA 01801
(617) 935-6460
Peanut Butter, Peanut Sauce

Upside Down Cajun Pickles 220
3421 N. Causeway Blvd.
Ste. 201
Metairie, LA 70002
(504) 834-5511

Van Douglas 55

Varda International Corp. 49
57-13 59th Street
Maspeth, NY 11378
(718) 821-9900

Vermillion Red 124

Victorian Victuals 71
809 Washington Street
Oregon City, OR 97045
(503) 650-0840
Victorian Tea Biscuits, Tea, Bread Mix

Viero s.r.l. 86, 87
Via Caselle
60-35010 Marsango (PD), Italy
(049) 552051-552014
Viero Savoiardi

Villa Nicola Ltd. 168
1616 Howard Street
Detroit, MI 48216
(313) 963-6542
Villa Nicola Oils

Voyageur Trading Co. 88, 89
Box 35121
Minneapolis, MN 55435
(612) 942-7766
(800) 445-7643
Voyageur Wild Rice

Wagner's Teas 126

Walker's Shortbread Ltd. 67
Aberlour-on-Spey
Scotland AB3 9PB
(034) 05555

Well Done Chips 120

W.J. Clark & Company 231, 247
355 North Ashland Ave.
Chicago, IL 60607
(312) 421-3676
(800) 383-3330
Woodland Pantry, Bean Cuisine Soups

World of Spices, Inc. 204
328 Essex Street
Sterling, NJ 07980
(201) 647-1218

Yuppie Gourmet, Inc. 14, 15, 16, 17
1419 Washington Ave.
P.O. Box 873
Racine, WI 53403
(414) 634-1110
Puffs Au Chocolate, Au Praline, Stix Au Chocolat